D1420440

CLARENDON ARISTOTLE SERIES

General Editor: J. L. ACKRILL

CLARENDON ARISTOTLE SERIES

POLITICS, Books III and IV

RICHARD ROBINSON

CATEGORIES and *DE INTERPRETATIONE*

J. L. ACKRILL

DE ANIMA, Books II and III

D. W. HAMLYN

ARISTOTLE'S
PHYSICS

BOOKS I AND II

Translated with Introduction and Notes

by

W. CHARLTON
LECTURER IN PHILOSOPHY
UNIVERSITY OF NEWCASTLE UPON TYNE

OXFORD
AT THE CLARENDON PRESS
1970

Oxford University Press, Ely House, London W. 1

GLASGOW NEW YORK TORONTO MELBOURNE WELLINGTON
CAPE TOWN SALISBURY IBADAN NAIROBI DAR ES SALAAM LUSAKA ADDIS ABABA
BOMBAY CALCUTTA MADRAS KARACHI LAHORE DACCA
KUALA LUMPUR SINGAPORE HONG KONG TOKYO

PRINTED IN GREAT BRITAIN

PREFACE

THE first aim of this as of other volumes in the Clarendon Aristotle series is to provide a translation of Aristotle's text sufficiently accurate to be used by serious students who know no Greek. The text used is that of W. D. Ross, *Aristotle's Physics, a revised text with introduction and commentary*, Oxford University Press, 1955. Words which Ross encloses in square brackets have been omitted. Departures from Ross's text, and points at which the translation seems to me uncertain, are marked with an asterisk and discussed in the *Notes on the text and translation*. The *Commentary* is addressed primarily to readers with some knowledge of philosophy, and intended to suggest starting points for the discussion of the philosophical value of Aristotle's ideas.

My gratitude is due in the first place to Prof. J. L. Ackrill, who read my drafts with great care, pointed out many errors, and made many helpful and stimulating suggestions. I should like also to acknowledge the encouragement of Prof. D. J. Allan, without whom this work would not have been undertaken. For most of the time I was engaged on it I was at Trinity College, Dublin, and much profited from discussions with my colleagues there. Finally, Mr. C. Kirwan has kindly shown me the part of his forthcoming volume in this series which deals with a chapter common to our two texts.

W. CHARLTON

Newcastle upon Tyne

CONTENTS

INTRODUCTION

THE first two books of Aristotle's *Physics* do not deal with problems in what we today call physics: Aristotle's own titles for them were probably 'Concerning principles' and 'Concerning nature' (Ross, pp. 1–6), and he writes as a philosopher, not as a scientist. Nevertheless, *Phys.* II, at least, seems to be addressed to the scientific student of nature (the *phusikos*: 194ᵃ16, ᵇ10, 198ᵃ22), and both books may, perhaps, most aptly be classified as philosophy of science. This seems to be roughly how Aristotle himself conceived them, though his demarcation of fields and methods of inquiry is tentative, and may appear a little strange and academic to the modern reader.

The student of nature deals with things which are subject to change (*Met. E* 1026ᵃ12), things which are not without matter (1026ᵃ6), things which have in themselves the source of their changing or staying unchanged (1025ᵇ20–1)—expressions at which we will look closely when we come to *Phys.* II. 1–2. Any question about such things Aristotle would call a 'physical' question (cf. *Top.* I. 105ᵇ19–29), but it does not follow that any discussion of such a question must be in every sense a 'physical' discussion.

The student of nature in the strictest sense, what we might call the natural scientist, bases his discussion of such questions on 'appropriate' premisses, that is, on principles which hold for physical things, things subject to change and so on, *as such*; and a discussion so based is 'physical' in the strict sense. In *De incess.* 704ᵇ12–24 Aristotle lists some assumptions from which people proceed when pursuing a physical method: that teleological explanation is valid in zoology, that there are six directions (up, down, right, left, in front, behind), that the source of locomotion is pushing and pulling. Similarly Democritus is said (*De gen. et cor.* I. 316ᵃ13) to have used physical and appropriate arguments, presumably because he argued from the hypothesis that things consist of atoms with primary qualities only, a hypothesis which, whether correct or incorrect, is appropriate to the topics under discussion (315^{a}34–^{b}9).

In *Phys.* I–II Aristotle is concerned with things which are subject to change, and hence with physical questions. He does not, however, pursue a physical method. So far from arguing from principles which hold for physical things as such, he is arguing to them: thus in *Phys.* II. 8 he is trying to establish the principle, mentioned in *De incessu*, of the validity of teleological explanation. How, then, did he conceive his method?

In *De gen. et cor.* I. 316^{a}11 Democritus is said to have proceeded 'physically' where Plato proceeded 'logically', and Aristotle might have called his procedure in our books 'logical'. He uses the word 'logical' (*logikos*) with a variety of nuances, but by a 'logical' argument he usually understands one proceeding from considerations which are not proper to the things being discussed. In *De gen. an.* II. 747^{b}28–30 he says of an argument: 'I call it logical, because in so far as it is more general, it is further from the appropriate principles.' A 'logical' argument is bad if the considerations on which it is based are not merely not appropriate to the subject under discussion, but appropriate to some other subject. Plato's argument in *De gen. et cor.* had that defect: it proceeded from considerations appropriate to geometry. But otherwise a logical argument may be acceptable or even necessary. We are told in *E.E.* I. 1217^{b}16–17 that a proper examination of Plato's views on the good would have to be logical, not ethical (cf. also *E.N.* I. 1096^{b}30–1). And Aristotle introduces his account of substance as form in *Met.* Z (1029^{b}13) with some 'logical' points. We might think that when it is a question of establishing 'appropriate' principles, logical argument is just what is needed.

Still, since the word 'logical' rather indicates what method is not pursued than what method is, Aristotle would probably not have called his method in *Phys.* I–II logical, but rather dialectical. Characterizing dialectic in *Soph. elench.* 11, he says that all disciplines make use of certain 'common things' (172ᵃ29); for the layman thinks he can challenge the expert up to a point, and that is, in so far as the expert is dealing with these common things (ibid. 31–2). Whilst we might wish to have these common things described more explicitly, Aristotle's idea is clearly that the expert's subject-matter has a side with which the expert's special knowledge, his 'appropriate principles', do not especially equip him to deal. Dialectic, we are told, deals in a technical or professional manner with this side of things, or these common things, with which others deal unprofessionally (ibid. 34–5). So that, although it has no determinate field (ibid. 12) in the way in which medicine and geometry have determinate fields (cf. 170ᵃ32–4), it is still a genuine discipline. This seems quite an apt description of the method of *Phys.* I–II: Aristotle is dealing in a technical manner with that side of the study of nature with which the natural scientist is not equipped to deal. Further, the special technique of the dialectician is to argue from *endoxa* (*Top.* I. 100ᵃ18–20), which are, roughly speaking, propositions which cannot be proved, but which an opponent could not deny without seeming unreasonable, and this is Aristotle's technique in *Phys.* I–II: he constantly appeals to what is ordinarily said or thought (e.g. 192ᵇ11–12, 194ᵇ33–5, 196ᵃ15–16, 199ᵃ1; see also below, pp. xv–xvi, for this aspect of his method); though he relies more on detailed linguistic analysis (e.g. 189ᵇ32–190ᵃ13) than the *Topics* might lead us to expect.

The method of the dialectician is the same as the method of the philosopher, except that the former uses it to win debates and the latter to ascertain the truth (*Top.* VIII. 155ᵇ7–10, I. 105ᵇ30, *Met.* *Γ* 1004ᵇ22–6). This suggests, since Aristotle in our books is presumably trying to ascertain the truth, that he would call them essays in philosophy, and in fact discussions of principles and causes parallel to those of *Phys.* I. 5–9 and II. 3 and 7 are found in books of the *Metaphysics* which are clearly conceived as philosophy (*philosophia* or *sophia*, *Met.* *A* 982ᵃ2 etc.). We might say, then, that in *Phys.* I–II the arguments are logical, the method is dialectic, and the discussions are philosophical; though this is perhaps misleadingly neat. As G. E. L. Owen (I. Düring and G. E. L. Owen, *Plato and Aristotle in the Mid-Fourth Century*, p. 164) suggests, we may doubt whether Aristotle when first composing the *Topics* recognized such a subject as *sophia* or philosophy over and above dialectical discussions of physical, logical, and ethical questions. It may have been only when he discovered that the 'common things' considered by the dialectician included forms of reasoning which could be separated off as the subject-matter of formal logic (or 'analytic', as he called it), that he made other 'logical' questions the province of a special subject.

Aristotle occasionally (e.g. *Met. Z* 1037ᵃ15) speaks of the philosophical discussion of things subject to change as 'second philosophy', by contrast with 'first philosophy', which is the philosophical study of things which are unchangeable. Not too much, however, should be made of this, for the unchangeable things which are the main topics of first philosophy are Platonic ideas and numbers, entities which Aristotle thinks do not exist. See the beginning of *Met. M*: we have now dealt, says Aristotle, with perceptible realities, and must see whether there is any kind of reality over and above them; we will begin by considering the opinions of others, of which there are two: some say that there are objects of mathematics, some that there are ideas (1076ᵃ8–19, cf. *B* 997ᵃ34–ᵇ3 etc.). Aristotle does indeed himself recognize another sort of unchangeable thing, the intelligent being which is the unchangeable source of change in the universe; but the discussion of this being he tends to call theology (*Met. E* 1026ᵃ19), and first philosophy for Aristotle stands to second philosophy much as the Dialectic in Kant's *Critique of Pure Reason* stands to the Analytic: as developed in *Met. M–N*, it is the exposure of the illusions of pure reason in its hyper-physical employment, and for Aristotle's positive and constructive philosophical teaching we must look to second philosophy.

Phys. I–II contain the formal introduction of a number of the basic concepts in Aristotle's philosophy: the matter–form distinction, the fourfold classification of causes, nature, and finality. For this reason, and because we are referred back to them by *Met. A* (983ᵃ33 f., 986ᵇ30–1, etc.), generally held to be an early work, an early date of composition has been assigned to them. Thus, according to Ross, 'we may say with some confidence that these two books were composed while Aristotle was still a member of the Academy' (p. 7). On the other hand, precisely because they seem to constitute the natural introduction to his other surviving works, we may think that as they stand—though they may incorporate the fruits of early speculation (M. Untersteiner suggests that *Phys.* I. 8–9 are taken from the early *De philosophia*)—they are the notes for lectures which were being delivered up to the end of Aristotle's career. How else did the student who entered the Lyceum make his way into Aristotle's system, if not through them? And if they were the regular first course in Aristotelian philosophy, presumably they were constantly revised and kept up to date. Such a presumption is supported by the sophistication of much of the argument, by the confident way in which Aristotle writes, as if he had a large and fully articulated body of material in reserve, and by the coherence of what he says here with what he says elsewhere. I shall make free use of the *Metaphysics*, *De anima*, etc., to bring out the significance of passages in *Phys.* I–II, and I do not think there is any passage in these books which can most easily be understood as the expression of a view later corrected or discarded.

Phys. I–II rather complement one another than form a continuous treatise. Book I, however, with its emphasis on the constituents of physical things generally, is more about the philosophy of physics, whilst the second book, with its emphasis on the development of plants and animals, is more about the philosophy of the biological sciences.

Phys. I centres round a question which Aristotle says elsewhere (*Met. Z* 1028ᵇ2–4) always has been, still is, and always will be, the focus of inquiry and perplexity, and the Greek for which is *ti to on*. This is sometimes translated 'What is being?', but that would be a better translation of the more sophisticated formulation which Aristotle suggests we substitute for it, *tis hē ousia* (ibid.). *Ti to on* itself is a much vaguer question, something like 'What is there?', 'What exists?', 'What is real?', 'What is the world?'

As such, it can be handled in various ways. It can, for instance, be treated as a scientific question, as a demand for the most basic kind of stuff in the universe, for the ultimate constituents of matter. Or it can be treated as a philosophic question, as a demand for an account of how we use words like 'real' and 'exist', of what we mean by a thing, and so on—accounts which Aristotle tries to give in the *Metaphysics*. In *Phys*. I Aristotle takes an intermediate line. His search for 'principles' is a search for the *logically* distinguishable factors which must be acknowledged in a world pervaded by change and becoming. He is asking, 'What must there be if there is coming to be, passing away, and alteration?', and he replies by giving a logical or philosophical analysis of coming to be.

This approach is of considerable historical interest. The Presocratic physicists had not disentangled the scientific and philosophical issues in the question 'What exists?', and their failure to deal with the latter had had (if we are to believe 191^{a}23–33, ^{b}30–3, etc.) bad effects on their handling of the former. By separating out this philosophical issue, and offering a detailed and purely philosophical treatment of it, Aristotle removed *a priori* inhibitions on empirical inquiry. (It should be recognized that the credit for so doing is not exclusively his; he is carrying on work the beginnings of which can be seen in Plato's *Phaedo*, especially 97–9.)

The main line of argument runs through chapters 1 and 4–7. Chapter 1 is introductory. In chapter 4 Aristotle reviews the theories of the Presocratic physicists, and distinguishes them into two groups, according as they make or do not make room for qualitative change. Having dismissed the second group with arguments which may seem a little cavalier, he obtains from the first a spring-board for his own account, which begins in chapter 5. In that chapter he presents the case for making the principles of any physical thing a pair of opposites; in chapter 6 he presents the case for saying that there must always be a third, additional factor; and in chapter 7 he argues that these two views can be reconciled, if we suppose that the basic elements of things are an underlying thing and a form. It is important to recognize (as W. Wieland has shown at length) that the distinction between underlying thing and form is not a presupposition of the whole discussion, but a conclusion to which Aristotle argues, and argues, moreover, not from metaphysical principles, but from linguistic considerations, by considering how we ordinarily talk.

The remaining chapters 2–3 and 8–9 may be accounted for by Aristotle's general methodology. Aristotle says that it is improper to inquire *what* a thing is, until you have established *that* it is, i.e. established that there is such a thing (e.g. *An. Po.* II. 93^a19–20, but cf. *Met. E* 1025^b17), and his practice in the *Physics* reflects this view: thus with chance, the infinite, place, void, time, and cf. on nature at 193^a3 ff. Now chapter 7, which is the kernel of *Phys.* I, is in fact an analysis of becoming; according to his principles, then, Aristotle ought to show that there is such a thing as becoming, that things do come to be. Chapters 2–3 fill this need. In them Aristotle does not indeed try to prove that becoming is possible: that, he says in 185^a12–13, is something we assume; but he does try to refute the arguments of the Eleatic monists, who were the chief opponents of the possibility of becoming. These chapters, then, may be seen, not only as part of the review of Presocratic opinions on what exists, but as an attempt to show that the considerations which led people to do away with change and becoming are ill-grounded.

Chapters 8–9 also accord with Aristotle's ideas of how a philosophical exposition should proceed. When discussing the notion of place in *Physics* IV, he first enumerates the generally held opinions about place, and then goes on: 'We must try to carry out our elucidation of the nature of place in such a way that the problems are resolved, that what is generally thought to be true of place remains true of it, and that the cause of the awkwardness of place, and of the difficulties felt over it, is made clear. That is the most stylish mode of philosophical exposition' (211^{a}7–11; cf. *E.N.* VII 1145^{b}2–7). He follows this course elsewhere in the *Physics* (thus over change: Book III, chapter 2, 'this is why change is difficult to get hold of', etc.), and is obviously doing the same in Book I, chapters 8 and 9. He begins chapter 8 by saying 'We must now show that only if our analysis is accepted can the difficulties felt by our predecessors be removed', and in chapter 9 he is mainly showing where the Academy went wrong.

The topic of *Phys.* II might be said to be explanation in natural science: in chapters 3 and 7 Aristotle presents his celebrated fourfold classification of causes, which is in fact a classification of modes of explanation or types of explanatory factor, and in chapters 4–6 he tries to show how chance or luck can be fitted into it (196^{b}8–9). In the discussion of explanation generally, however, one issue stands out with special prominence, the validity of teleological explanation.

Chapter 1 begins with a distinction between natural objects and things like artefacts which are not due to nature. Natural objects are said to have a source of their behaviour in themselves, and nature is defined as such a source. Aristotle then goes on to claim that of the two factors in any physical thing distinguished in *Phys.* I, matter and form, not only the first but the second also can be its nature in this sense. This thesis is tackled from various angles in chapters 1, 2, 8, and 9, and most formally in chapter 8, where it is represented as the thesis that 'nature is a cause *for* something' (198^{b}10–11), i.e. that some natural things and processes exist or come about for the sake of definite ends, and can be explained as existing and coming about for those ends.

If the argument in these chapters is to be followed, three points, as I shall try to show in detail in the commentary, must be kept in mind. First, when Aristotle talks about nature, he is not talking about a single universal force, which pervades all natural objects and directs their development and behaviour towards goals it has appointed for them. There are passages in his works (e.g. *De caelo* II. 291ᵃ24–6, *De part. an.* IV. 687ᵃ10–12) which might suggest a belief in such a force, but it is usually and, I think, rightly judged that they are figurative, or at most betray a privately held theological opinion (cf. *De caelo* I. 271ᵃ34, *De gen. et cor.* II. 336ᵇ27–32). When he is writing as a scientist or as a philosopher of science he means by nature the nature of this or that thing. We say that a natural object, like a tree or a horse, has a nature: it is that nature which it has, which in *Phys.* II Aristotle is trying to get at. Second, for Aristotle the question whether something can or cannot be explained teleologically, as being '*for* something', is equivalent to the question whether, in its case, matter or form is nature in the sense of source of its coming to be. Aristotle would not contrast explanation by final with explanation by formal causes, at least within the field of natural history: for him it is obvious that if the form of a plant or animal explains its behaviour, it explains it as final cause; and conversely, if it is correct to say that a tiger's teeth are for biting and its stripes for camouflage, that is as much as, and no more than, to say they are accounted for by the form of the tiger, not by its matter. Third, Aristotle does not argue that everything which is due to nature is due to form and susceptible of teleological explanation. He proposes teleological explanations only in cases where it seems correct to speak of some form of life. This does not emerge too clearly from his writings, because he devotes (not in *Phys.* II but elsewhere) much space to the heavenly bodies, and leans (but with some ambiguity) to the speculation that they are alive (*De caelo* II. 285ᵃ27–31, 292ᵃ18–21); so that sometimes their behaviour is attributed to the stuff of which they are made, sometimes to a Deity which moves them as an object of thought and desire. When, however, as in the *Meteorologica*, he deals with sublunary physical phenomena, such as weather, the sea, coction, his explanations are exclusively in terms of necessity, chance, and the natures of different kinds of matter.

BOOK I

CHAPTER 1

In all disciplines in which there is systematic knowledge of 184a things with principles, causes, or elements, it arises from a grasp of those: we think we have knowledge of a thing when we have found its primary causes and principles, and followed it back to its elements. Clearly, then, systematic knowledge of nature must start with an attempt to settle questions about principles.

The natural course is to proceed from what is clearer and more knowable to us, to what is more knowable and clear by nature; for the two are not the same. Hence we must start thus with things which are less clear by nature, but clearer to us, and move on to things which are by nature clearer and more knowable. The things which are in the first instance clear and plain to us are rather those which are compounded. It is only later, through an analysis of these, that we come to know elements and principles.

That is why we should proceed from the universal to the particular. It is the whole which is more knowable by perception, and the universal is a sort of whole: it embraces many things as parts. Words stand in a somewhat similar 184b relationship to accounts. A word like 'circle' indicates a whole indiscriminately, whereas the definition of a circle divides it into particulars. And little children at first call all men father and all women mother, only later coming to discriminate each of them.

CHAPTER 2

There must be either one principle or more than one. If one, it must be either unchangeable, the view of Parmenides and Melissus, or subject to change, the view of the physicists, of whom some make air and others water the primary principle.

If there are more principles than one, they must be either limited in number—that is, there are either two, three, four, or some such definite number of them—or unlimited. In the latter case, either they are all the same in kind, and ⟨differ⟩ only in shape, as Democritus held, or they are different or even opposed in species. We are here raising the same question as those who ask how many things there are: they are really inquiring about the primary constituents of things, whether they are one or several, and if several, whether they are limited or unlimited in number, so they too are inquiring into the number of principles and elements.

Now the question whether what is is one and unchangeable, does not belong to a discussion of nature. Just as the geometer has nothing left to say to the man who does away with the principles of geometry, but must refer him to a student of something else, or of what is common to all studies, so it is when we are inquiring into principles: there will be no principle left if what is is one thing only, and one in this way. A principle must be a principle of some thing or things. Discussing whether what is is one in this way, is like discussing any other thesis advanced for the sake of having a discussion, like that of Heraclitus, or the view that what is is a single man. Or like exposing a quibble, such as is latent in the arguments of both Melissus and Parmenides: for both reason invalidly from false premisses, but Melissus is the duller and more obvious: grant him one absurdity and he is able to infer the rest—no great achievement.

For ourselves, we may take as a basic assumption, clear from a survey of particular cases, that natural things are some or all of them subject to change. And we should not try to expose all errors, but only those reached by arguing from the relevant principles; just as it is the geometer's job to refute a quadrature by means of lunes, but not one like Antipho's. Nevertheless, since, though they are not writing about nature, the Monists happen to raise difficulties pertinent to it, we would do well, perhaps, to say a little about them; for the inquiry offers scope for philosophy.

The most appropriate way of all to begin is to point out that things are said to be in many ways, and then ask in what way they mean that all things are one. Do they mean that there is nothing but reality, or nothing but quantity or quality? And do they mean that everything is one single reality, as it might be one single man, or one single horse, or one single soul, or, if all is quality, then one single quality, like pale,* or hot, or the like? These suggestions are all very different and untenable. If there is to be reality and quality and quantity, then whether these are apart from one another or not, there will be more things than one. And if everything is quality or quantity, then whether there is also reality or not, we run into absurdity, if, indeed, impossibility can be so called. Nothing can exist separately except a reality; everything else is said of a reality as underlying thing.

Melissus says that what is is unlimited. It follows that what is is some quantity. For the unlimited is unlimited in quantity, and no reality, quality, or affection can be unlimited, except by virtue of concurrence, there being also certain quantitative things. For quantity comes into the account of the unlimited, but reality and quality do not. If, then, there is reality and quantity as well, what is is twofold and not one; if there is just reality, so far from being unlimited, it will have no magnitude at all; if it had, there would be some quantity.

Again, as things are said to be, so they are said to be one, in many ways; so let us see in what way the universe is supposed to be one. A thing is called one if it is a continuum, or if it is indivisible, and we also call things one if one and the same account is given of what the being of each would be: so, for instance, wine and the grape.

Now if the universe is continuous, the one will be many; for continua are divisible without limit. (There is a difficulty about parts and wholes, though perhaps it is a problem on its own and not relevant to the present discussion: are the parts and the whole one thing or several, and in what way are they one or several, and if several, in what way are they several? And what about the parts which are not continuous?

And is each indivisibly one with the whole, since they will be the same with themselves also?)

Is the universe one, then, in that it is indivisible? Then nothing will have any quantity or quality, and what is will be neither unlimited, as Melissus says, nor limited, as Parmenides prefers. For it is limits which are indivisible, not limited things.

If, however, all things are one in account, like raiment and apparel, they will find themselves in the position of Heraclitus. The being of good and the being of bad, of good and not good, will be the same, so that good and not good, man and horse, will be the same, and the thesis under discussion will no longer be that all things are one, but that they are nothing at all. And the being of a certain quality and the being of a certain quantity will be the same.

Thinkers of the more recent past also were much agitated lest things might turn out to be both one and many at the same time. Some, like Lycophron, did away with the word 'is'; others sought to remodel the language, and replace 'That man is pale' 'That man is walking', by 'That man pales' 'That man walks', for fear that by inserting 'is' they would render the one many—as if things were said to be or be one in only one way. Things, however, are many, either in account (as the being of pale is different from the being of a musician*, though the same thing may be both: so the one is many), or by division, like the parts of a whole. At this point they got stuck, and began to admit that the one was many; as if it were not possible for the same thing to be both one and many, so long as the two are not opposed: a thing can be one in possibility and in actuality.

CHAPTER 3

If we approach the matter thus, it appears to be impossible that things are all one, and the arguments in fact adduced are not hard to rebut. Both of them, Melissus and Parmenides, argue in quibbles; they reason invalidly from false premisses;

but Melissus is the duller and more obvious: grant him one absurdity, and he is able to infer the rest—no great achievement.*

The fallacies of Melissus are patent. He thinks that if he has made it a premiss that whatever comes to be has a beginning, he has also made it a premiss that whatever does not come to be has no beginning. It is also absurd to say that in all cases there is a beginning, not only of the time, but of the thing, and that, not only when the coming to be is a coming simply into being, but also when it is a qualitative change—as if change never took place on an extended front. And then, how does it follow, because all is one, that all is unchangeable? If a part of the universe which is one, like this water here, can change in itself, why not the whole? And why should there be no such thing as qualitative change? In fact, the contents of the universe cannot be one even in species—men and horses are different in species and so are opposites—unless inasmuch as they are made of the same sort of stuff; and some of the physicists, indeed, say that all is one in that way, though not in the other.

Parmenides is open to all these objections, besides others exclusive to himself. The answer to him is that he assumes what is not true and infers what does not follow. His false assumption is that things are said to be in only one way, when they are said to be in many. As for the invalidity, suppose we say that there are only pale things, and that 'pale' means only one thing: the pale things will be none the less many and not just one. The pale will not be one in virtue of being continuous, nor will it be one in account. For the being of pale will be different from the being of that which has received it. By that I do not imply that anything can exist separately except the pale: it is not because they can exist separately, but because they differ in their being, that the pale and that to which it belongs are different. This, however, is something Parmenides did not get far enough to see.

He must make it a premiss, then, not only that 'is' means

only one thing, whatever is said to be, but that it means precisely what is, and precisely what is one. For that which supervenes is said of some underlying thing, so if 'is' supervenes, that on which it supervenes will not be, for it will be something different from that which is; and therefore there will be somewhich is not. Precisely what is, then, will not be something which belongs to something else. It cannot be a particular sort of thing which is, unless 'is' means more things than one, such that each is a sort of being, and it was laid down that 'is' means only one thing.

But now, if precisely what is does not supervene on anything else, but ⟨other things⟩ rather supervene on it, why does 'precisely what is' mean 'is' more than 'is not'? Suppose that precisely what is is also pale, and that the being of pale is not precisely what is (for being cannot even supervene on it, since nothing is a thing which is except precisely what is): it will follow that that which is pale is not. And I do not mean that it will not be this or that: it will not be at all. But then precisely what is will not be: for it was true to say that it was pale, and that meant something which is not. So 'pale' also must mean precisely what is. But then 'is' will have more than one meaning.

Again, if what is is precisely what is, then what is will not have magnitude, for the being of each of its parts would be different.

That precisely what is divides into something else which is precisely what is, is clear as soon as we try to give an account. Suppose a man is* precisely what is; then animal must be something which is precisely what is, and so must biped. If not, they must be supervenient; must supervene, then, either on man or on some other underlying thing; and neither alternative will stand.

A thing is called supervenient, either if it is such that it can belong or not belong* [or if that on which it supervenes comes into the account of it] or if the account of that on which it supervenes comes into it. Thus being seated is supervenient in that it is separable, and the account of the nose on which we say snub supervenes comes into snub. Further, whatever enters as a constituent into the definitory account of a thing must be such that the account of the whole thing does not enter into the account of it. Thus the account of man does not come into biped, and the account of pale man does not come into pale. That being so, if biped supervened on man, either biped would have to be separate from man, so that we could have men who were not bipeds; or the account of man would enter into the account of biped. This last, however, is impossible, since biped comes into the account of man.

If, on the other hand, animal and biped supervene on something other than man, and each is not something which is precisely what is, man too will be something which supervenes on something else. But we must take it that precisely what is does not supervene on anything, and if both [and each] of two things are said of something, so must that which they constitute. Does the universe, then, consist of indivisibles?

Some people gave in to both arguments: to the argument that if 'is' means only one thing, all things must be one, when they said that there is that which is not; and to the argument from dichotomy, when they posited indivisible magnitudes. But it is obviously untrue that if 'is' means only one thing, and nothing can both be and not be, there will be nothing which is not. That which is not, need not not be* altogether: it may not be something definite. And to say that if nothing is over and above what is itself, all things will be one, is absurd. For who understands by 'what is itself' anything but precisely what is something definite? If that is so, there is still nothing to stop there being a plurality of things, as has been said. That it is impossible, then, for what is to be one in the way claimed is clear.

CHAPTER 4

There are two main lines taken by the physicists. Some make the underlying body one, making it either one of the three, or something else which is more solid than fire but less solid than air. From this they produce everything else (and they allow a plurality of products) by means of density and rarity. These are opposites, and, to put it in general terms, are excess and defect, as are the great and small in Plato; though Plato differs from them in that he makes the great and small matter, and what is one the form, whilst they make the one underlying thing matter, and the opposites differentiating principles and forms.

The other line is taken by those who, like Anaximander, make the one stuff already contain in it oppositions, which are then separated out, and also by those who say that it is both one and many, like Empedocles and Anaxagoras. They too posit a hotchpotch, from which everything else emerges by separation. They differ, however, in that Empedocles posits periodic mixtures and separations, whilst Anaxagoras posits only one, and Anaxagoras posits an unlimited number both of homeomerous elements and of opposites, while Empedocles posits only the elements which are so called.*

Anaxagoras probably made his elements unlimited in this way because he accepted as true the general opinion of the physicists that nothing comes to be out of what is not. It is on this ground that they say that things were once 'all together', and that he makes the coming to be of a thing of a certain sort alteration, while they make it coming together and dissolution. It was also a consideration, that opposites come to be out of one another: they must, it seemed, have been there all the time. For if everything which comes to be must do so either out of what is or out of what is not, then, if the latter is impossible (and about that there is unanimity among all who discuss nature), the former, they thought, must be true: everything comes to be out of things which already exist and are present, but cannot be perceived by us
187b because they are extremely tiny. According to them, then, everything is mixed together in everything, because they saw everything coming to be out of everything: things only look different, and are said to be one thing rather than another, because there is a numerical preponderance of that in the

mixture of the unlimited particles; there is no whole object which is purely pale, dark, sweet, flesh, or bone, but whichever a thing has most of is commonly taken as constituting its nature.

Now if the unlimited as such is unknowable, then there is no knowing the quantity of that which is unlimited in number or size, and no knowing what sort of thing a thing is, if there is no limit to its forms. If, then, the principles are unlimited both in number and in forms, there can be no knowledge of the things they make up. For we think we have knowledge of something composite, when we know the variety and number of its components.

Further, if it is necessary that, if a part of a thing (and I am speaking of the parts into which, as constituents present in it, the whole can be divided) can be as large or small as you please, then so can the whole, and if it is not possible for any animal or plant to be as large or small as you please, it is not possible that any part should be either; for if it could, so could the whole. Now flesh and bone are parts of animals, and fruits are parts of plants. Clearly, then, neither flesh nor bone nor anything of that sort can proceed indefinitely far either in enlargement or in diminution.

Again, if all such things are already present in one another, and do not come into existence, but are merely separated out after being there all along, objects getting their appellation from whatever is present in most abundance; and if anything can come to be out of anything, for instance water be separated out from flesh, and flesh from water; and if only a limited quantity of stuff is needed to do away with a limited quantity of stuff: it plainly follows that everything cannot be present in everything. For suppose that some flesh is removed from some water, and then more flesh extracted from what remains: even if the yield is lower each time, there will still always be some quantity smaller than any yet yielded. Hence either the separating out will come to an end, in which case the residue of water will be completely void of flesh, and it will not be true that everything is in everything; or else it will not come to an end, but more can always be extracted, in which case we shall have the impossibility that there is an unlimited number of equal limited parts in a limited magnitude.

Further, if every body must become smaller when something is removed from it, and if flesh cannot increase or diminish in quantity beyond a certain limit, it is plain that from the least
188a possible quantity of flesh nothing corporeal can be extracted. For there would then be a quantity of flesh smaller than the least possible.

Again, the unlimited number of corporeal particles would each contain an unlimited supply of flesh, blood, and brain, ⟨not⟩ indeed separated from one another, but none the less real and unlimited; but that is nonsense.

That the separating out will never be complete is true, but Anaxagoras did not understand why. Affections are not capable of separation. If colours and states are mixed together, then if they get separated out, we shall have a pale or a healthy which is nothing else, which is not even *of* an underlying thing. Hence Anaxagoras' Mind is absurd: it is seeking the impossible, since it wants to effect a separating which cannot be effected, whether it is conceived as a separation of quantities or of qualities: from the former angle, because there is no smallest magnitude, from the latter, because affections are not capable of separation. Anaxagoras did not get right even the coming to be of things of the same species, for in one way clay divides into clay but in another it does not. As for his suggestion that water and air are constituted and come to be out of one another in the way in which you get bricks out of houses and houses out of bricks, the cases are not parallel. Altogether it is better to make your basic things fewer and limited, like Empedocles.

CHAPTER 5

That opposites are principles is universally agreed: by those who say that the universe is one and unchanging (for Parmenides in effect makes hot and cold principles, though he calls them fire and earth), by those who make use of dense and rare, and by Democritus. Democritus posits the full and the empty, saying that the one is present as that which is, and the other as that which is not; he also makes use of position, shape, and arrangement, and these are genera of opposites: position comprises above and below, in front and behind; shape comprises angular and smooth, straight and curved. Clearly, then, all in some way agree that opposites are the principles. And that is plausible. For the principles must come* neither from one another nor from anything else, and everything else must come from them. Primary opposites fulfil these conditions: because they are primary they do not come from anything else, and because they are opposite they do not come from one another. But we must also see what emerges from logical considerations.

Our first point must be that nothing whatever is by nature such as to do or undergo any chance thing through the agency of any chance thing, nor does anything come to be out of just anything, unless you take a case of concurrence. For how could pale come to be out of knowing music, unless the knowing music supervenes on the not pale or the dark? Pale comes to be out of not pale—not, that is, out of just anything other than pale, but out of dark or something between the two; 188b
and knowing music comes to be out of not knowing music, that is, not out of just anything other than knowing music, but out of ignorant of music, or something in between if there is anything in between. And a thing does not pass away into just anything in the first instance; thus the pale does not pass away into the knowing music, except by virtue of concurrence, but into the not pale, and not into any chance thing other than the pale, but into the dark or something in between. Similarly the knowing music passes away into the not knowing music, and not into any chance thing other than the knowing music, but into the ignorant of music or something in between if there is anything in between. It is the same in all other cases, since the same account holds for things which are not simple but composite, though we do not notice, because the opposed

dispositions have no name. It is necessary that the united* should always come to be out of disunited, and the disunited out of united, and that the united should pass away into disunion, and not just any chance disunion, but that opposed to the preceding union. And it makes no difference whether we speak of union, or arrangement, or composition: plainly the same account holds. And a house, a statue, what you please, comes to be in the same way. The house comes to be out of the not being put together but dispersed thus of these materials, and the statue or anything else which is shaped, arises out of shapelessness. Every one of these things is an arrangement or composition.

If this is true, everything which comes to be comes to be out of, and everything which passes away passes away into, its opposite or something in between. And the things in between come out of the opposites—thus colours come out of pale and dark. So the things which come to be naturally all are or are out of opposites.

So far most thinkers are prepared to go along with us, as I said above. For they all represent their elements and what they call their principles as opposites, even if they give no reason for doing so, as though the truth itself were forcing them on. They differ among themselves in that some take pairs which are prior and some take pairs which are posterior, and some choose pairs which are more readily known with the aid of an account, and some choose pairs which are more readily known by perception: for some put forward hot and cold as the causes of coming to be, and others wet and dry, and others odd and even or strife and love, and these differ in the manner just stated. So the pairs they propose are in a way the same and different: they are different, as indeed they are generally thought to be, but by analogy the same. For they are taken from the same list, some of the opposites being wider in extent and others included under them. This is how it is that the principles put forward are the same and different, and some better and some worse; and some, as we said, more easily known with the aid of an account, like the great and the small, and others more easily known by perception, like the rare and the dense—for that which is universal is more easily known in the former way, since accounts are of what is universal, and that which is particular in the latter, since perception is of particulars.

That the principles, then, must be opposites is plain.

CHAPTER 6

We must next say whether they are two, three, or more in number.

They cannot be one, since opposites are not one and the same; and they cannot be unlimited, since if they were, what is would be unknowable, since there is one opposition in any one kind of thing, and reality is one such kind, and since we can get on with a limited number, and it is better to use a limited, like Empedocles, than an unlimited. Empedocles claims to do everything Anaxagoras can do with his unlimited plurality. Further, some pairs of opposites are prior to others, and some, like sweet and bitter, pale and dark, arise from others,* whereas principles ought to be constant.

That shows that they can be neither one nor unlimited in number. But if they are limited, there is an argument for not making them only two. For it is hard to see how density could be by nature such as to act on rarity or vice versa, and similarly whatever the opposition: love does not gather up strife and make something out of it, nor does strife act thus with love, but both must act on a third thing distinct from them. And some people enlist even more principles to constitute the nature of things.

We may also run into the following difficulty if we do not posit some additional nature to underlie the opposites. We never see opposites serving as the reality of anything, and yet a principle ought not to be something said of some underlying thing. If it is, the principle will itself have a principle, for that which underlies is a principle, and is thought to be prior to that which is said of it.

Again, we do not say that one reality is the opposite of another. How, then, can a reality be constituted by things which are not realities? And how can that which is not a reality be prior to that which is?

Anyone, then, who accepts both the earlier argument and this, must, if he is to preserve both, posit some third thing which underlies, as do those who say that the universe is one single nature, such as water or fire or something between the two. The last suggestion is the most hopeful, since fire, earth, air, and water are already tangled up with oppositions. Those, then, are not without reason, who make the underlying thing different from any of these, or, if one of them, air, since that has the least perceptible differentiating features. After it comes water. Anyhow, all shape their one stuff with the opposites, with density and rarity and the more and the less; and these clearly, as I said above, are, in general terms, excess and defect. It does not seem to be at all a novel idea, that the principles of things are the one, excess and defect, though it has been put forward in different ways: earlier inquirers made the single principle passive and the pair active, whilst certain more recent thinkers prefer to turn it round and say that it is the one which is active and the pair passive.

That there are as many as three elements, then, may seem arguable to anyone guided by these and similar considerations; but at three we might draw the line. The single one is enough for being acted on; and if there are four, giving us two oppositions, we shall have to supply a further intermediate nature for each separately. Or if there are two pairs and they can produce things out of one another, one of the oppositions will be otiose. Moreover, there cannot be more than one primary opposition. Reality is a single kind of thing, so that the principles can differ only in being prior or posterior to one another, and not in kind. In any one kind there is always one opposition, and all oppositions seem to reduce to one.

That the elements, then, are neither one in number, nor more than two or three, is plain; but whether they are two or three is, as I have said, a very difficult question.

CHAPTER 7

This is how I tackle it myself. I shall be dealing first with coming to be in general, since the natural procedure is first to say what is common to all cases, and only then to consider the peculiarities of each.

When we say that one thing comes to be out of another, or that something comes to be out of something different, we may be talking either about what is simple or about what is compound. Let me explain. A man can come to be knowing music, and also the not knowing music can come to be knowing music, or the not knowing music man a man knowing music. I call the man and the not knowing music simple coming-to-be things, and the knowing music a simple thing which comes to be. When we say that the not knowing music man comes to be a knowing music man, both the coming-to-be thing and that which comes to be are compound.

In some of these cases we say, not just that this comes to be, but that this comes to be out of this—for instance, knowing music comes to be out of not knowing music. But not in all: knowing music does not come to be out of man, but the man comes to be knowing music.

Of what we call the simple coming-to-be things, one remains when it comes to be, and the other does not. The man remains and is a man when he comes to be knowing music,* but the not knowing music and the ignorant of music do not remain, either by themselves or as components.

These distinctions having been made, in all cases of coming to be, if they are looked at as I suggest, this may be taken as definite, that there must always be something underlying which is the coming-to-be thing, and this, even if it is one in number, is not one in form. (By 'in form' I mean the same as 'in account'.) The being of a man is not the same as the being of ignorant of music. And the one remains and the other does not. That which is not opposed remains—the man remains—but the not knowing music and the ignorant of music do not remain, and neither does the compound of the two, the ignorant of music man.

We say that something comes to be out of something, and not that something comes to be something, chiefly in connection with that which does not remain. Thus we say that knowing music comes to be out of not knowing music, but we do not say that it comes to be out of man. Though we sometimes speak thus about things that do remain: we say that a statue comes to be out of bronze, not that bronze comes to be a statue. But we speak in both ways of that which comes to be out of what is opposed to it and does not remain: we say both 'this comes to be out of this' and 'this comes to be this'. Out of ignorant of music comes to be knowing music, and ignorant of music comes to be knowing music. Hence it is the same with the compound; we say both that out of a man who is ignorant of music, and that a man who is ignorant of music, comes to be one who knows music.

Things are said to come to be in many ways, and some things are said, not to come to be, but to come to be something, while only realities are said simply to come to be. In the case of other things it is plain that there must be something underlying which is the coming-to-be thing—for when a quantity, quality, relation, [time,] or place comes to be, it is *of* an underlying thing, since it is only realities which are not said of anything further, and all other things are of realities. But that realities too, and whatever things simply are, come to be out of something underlying, will, if you look attentively, become plain. There is always something which underlies, out of which the thing comes to be, as plants and animals come to be out of seed. The things which simply come to be do so some of them by change of shape, like a statue, some by addition, like things which grow, some by subtraction, as a Hermes comes to be out of the stone, some by composition, like a house, some by alteration, like things which change in respect of their matter. All things which come to be like this plainly come to be out of underlying things.

From what has been said, then, it is clear that that which comes to be is always composite, and there is one thing which comes to be, and another which comes to be this, and the latter is twofold: either the underlying thing, or the thing which is opposed. By that which is opposed, I mean the ignorant of music, by that which underlies, the man; and shapelessness, formlessness, disarray are opposed, and the bronze, the stone, the gold underlie.

Plainly then if there are causes and principles of things which are due to nature, out of which they primarily are and have come to be not by virtue of concurrence, but each as we say when we give its reality,* everything comes to be out of the underlying thing and the form. For the knowing music man is composed in a way of man and knowing music. Analyses are into accounts of these two. So it is clear that things which come to be come to be out of them. The underlying thing, however, though one in number, is two in form. On the one hand there is the man, the gold, and in general the measurable matter; this is more of a this thing here, and it is not by virtue of concurrence that the thing which comes to be comes to be from this. On the other hand there is the lack or opposition, which is supervenient. As for the form, it is one: it is the arrangement, or the knowledge of music, or some other thing said of something in the same way. Hence from one angle we must say that the principles are two, and from another that they are three; and from one angle they are the opposites—as when we say that they are the knowing music and the ignorant of music, or the hot and the cold, or the united and the disunited—but from another angle not, for opposites cannot be acted upon by one another. This difficulty too is resolved by the fact that the underlying thing is something else, and that other thing is not an opposite. So in one way the principles are not more numerous than the opposites, but are, you might say, two in number; but

191a they are not two in every way, because of the diverse being which belongs to them, but three. (For the being of a man is different from the being of ignorant of music, and the being of shapeless from the being of bronze.)

How many principles there are of natural things [which are involved in coming to be],* and in what way they are so many, has now been said. It is clear that there must be something to underlie the opposites, and that the opposites must be two in number. Yet in another way that is not necessary. One of the opposites, by its absence and presence, will suffice to effect the change.

As for the underlying nature, it must be grasped by analogy. As bronze stands to a statue, or wood to a bed, or [the matter and] the formless before it acquires a form to anything else which has a definite form, so this stands to a reality, to a this thing here, to what is. This, then, is one principle, though it neither is, nor is one, in the same way as a this thing here; another principle is that of which we give the account; and there is also the opposite of this, the lack. In what way these principles are two, and in what way more than two, has been said above. The theory originally was that the only principles were the opposites; then that there had to be something else to underlie them, making the principles three; on our present showing it is plain what sort of opposites are involved,* how the principles stand to one another, and what the underlying thing is. Whether the form or the underlying thing has the better claim to be called the reality, is still obscure; but that the principles are three, and how, and what the manner of them is, is clear.

So much on how many and what the principles are.

CHAPTER 8

That this is the only way of resolving the difficulty felt by thinkers of earlier times must be our next point. The first people to philosophize about the nature and truth of things got so to speak side-tracked or driven off course by inexperience, and said that nothing comes to be or passes away, because whatever comes to be must do so either out of something which is, or out of something which is not, and neither is possible. What is cannot come to be, since it is already, and nothing can come to be out of what is not, since there must be something underlying. And thus inflating the consequences of this, they deny a plurality of things altogether, and say that there is nothing but 'what is itself'.

They embraced this opinion for the reasons given. We, on the other hand, say that it is in one way no different, that something should come to be out of what is or is not, or that what is or is not should act on or be acted on by something, or come to be any particular thing, than that a doctor should act on or be acted on by something, or that anything should be or come to be out of a doctor. By this last we may mean two things, so clearly it is the same when we say that something is out of something which is, and that what is acts or is acted on. A doctor builds a house, not as a doctor, but as a builder, and comes to be pale, not as a doctor, but as dark. But he doctors and comes to be ignorant of medicine as a doctor. Now we must properly say that a doctor acts or is acted on, or that something comes to be out of a doctor, only if it is as a doctor that he does or undergoes or comes to be this. So clearly to say that something comes to be out of what is not, is to say that it does so out of what is not, as something which is not. They gave up through failing to draw this distinction, and from that mistake passed to the greater one of supposing that nothing comes to be or, apart from what is itself, is, thus doing away with coming to be altogether. We too say that nothing comes to be simply out of what is not; but that things do come to be in a way out of what is not, sc. by virtue of concurrence. A thing can come to be out of the lack, which in itself is something which is not, and is not a constituent. This, however, makes people stare, and it is thought impossible that anything should come to be in this way, out of what is not.

Similarly there can be no coming to be out of what is or of what is, except by virtue of concurrence. In that way, however, this too can come about, just as if animal came to be out of animal and animal of a particular sort out of animal of a particular sort, for instance dog ⟨out of dog or horse⟩ out of horse. The dog would come to be, not only out of a particular sort of animal, but out of animal; not, however, as animal, for that belongs already. If a particular sort of animal is to come to be, not by virtue of concurrence, it will not be out of animal, and if a particular sort of thing which is, it will not be out of thing which is; nor out of thing which is not. We have already said what it means to say that something comes to be out of what is not: it means out of what is not, as something which is not. Further, there is no violation here of the principle that everything either is or is not.

That is one way of handling the matter; another is to point out that the same things may be spoken of either as possible or as actual. That, however, is dealt with in greater detail elsewhere.

So, as we have said, the difficulties are resolved, by which people were driven to do away with some of the things mentioned. For that was why the earlier thinkers too were diverted so far from the path to coming to be, passing away, and change generally; when this nature, if they had seen it, would have put them right.

CHAPTER 9

Others, indeed, have touched its surface, but they did not go deep enough. In the first place, they agree that it is in a general way the case that a thing comes to be out of what is
192a not,* and that so far Parmenides was right. And then it appears to them that if it is one in number, it is only one in possibility, which is not at all the same thing. We for our part say that matter and lack are different, and that the one, the matter, by virtue of concurrence is not, but is near to

reality and a reality in a way, whilst the other, the lack, in itself is not, and is not a reality at all. According to them, on the other hand, the great and the small, whether together or separate, are what is not in the same way. So their three things and ours are completely different. They got as far as seeing that there must be an underlying nature, but they made it one. And if someone calls it a pair, viz. great and small, he is still doing the same thing, for he overlooked the other nature. The one remains, joint cause with the form of the things which come to be, as it were a mother. The other half of the opposition you might often imagine, if you focus on its evil tendency, to be totally non-existent. Given that there is one thing which is divine and good and yearned for, our suggestion is that there is one thing which is opposite to this, and another which is by nature such as to yearn and reach out for it in accordance with its own nature. They, however, will find that the opposite is reaching out for its own destruction. But the truth is that neither can the form yearn for itself, since it is in need of nothing, nor can its opposite yearn for it, since opposites are mutually destructive, but it is the matter which does the yearning. You might say that it yearns as the female for the male and as the base for the beautiful; except that it is neither base nor female, except by virtue of concurrence.

And in one way it passes away and comes to be, and in another not. Considered as that in which, it does in itself pass away [for that which passes away, the lack, is in it].* Considered, however, as possible, it does not in itself pass away, but can neither be brought to be nor destroyed. If it came to be, there would have to be something underlying, out of which, as a constituent, it came to be; that, however, is the material nature itself, for by matter I mean that primary underlying thing in each case, out of which as a constituent and not by virtue of concurrence something comes to be; so it would have to be before it had come to be. And if it passed away, this is what it would ultimately arrive at, so it would have passed away before it had passed away.

As for the formal principle, whether such principles are one or many, and of what sort or sorts they are, are questions to be treated in detail in first philosophy, so we may leave them aside until we come to that. In what follows we shall be speaking of natural forms which can pass away.

That there are principles, then, and what and how many they are, we may take as settled in this way. Let us now proceed, making a fresh start.

BOOK II

CHAPTER 1

SOME things are due to nature; for others there are other causes. Of the former sort are animals and their parts, plants, and simple bodies like earth, fire, air, and water—for we say that these and things like them are due to nature. All these things plainly differ from things which are not constituted naturally: each has in itself a source of change and staying unchanged, whether in respect of place, or growth and decay, or alteration. A bed, on the other hand, or a coat, or anything else of that sort, considered as satisfying such a description, and in so far as it is the outcome of art, has no innate tendency to change, though considered as concurrently made of stone or earth or a mixture of the two, and in so far as it is such, it has. This suggests that nature is a sort of source and cause of change and remaining unchanged in that to which it belongs primarily and of itself, that is, not by virtue of concurrence. What do I mean by that qualification? Well, a man who is a doctor might come to be a cause of health in himself. Still, in so far as he is healed he does not possess the art of medicine, but being a doctor and being healed merely concur in the same person. Were the matter otherwise, the roles would not be separable.

Similarly with other things which are made. They none of them have in themselves the source of their making, but in some cases, such as that of a house or anything else made by human hands, the source is in something else and external, whilst in others the source is in the thing, but not in the thing of itself, i.e. when the thing comes to be a cause to itself by virtue of concurrence.

Nature, then, is what has been said, and anything which has a source of this sort, has a nature. Such a thing is always

a reality; for it is an underlying thing, and nature is always in an underlying thing. It is in accordance with nature, and so is anything which belongs to it of itself, as moving upwards
193a belongs to fire—for that neither is a nature nor has a nature, but is due to nature and in accordance with nature.

We have now said what nature is and what we mean by that which is due to nature and in accordance with nature. That there is such a thing as nature, it would be ridiculous to try to show; for it is plain that many things are of the sort just described. To show what is plain by what is obscure is a sign of inability to discriminate between what is self-evident and what is not—and it is certainly possible to be so placed: a man blind from birth would have to make inferences about colours. For such people discussion must be about the words only, and nothing is understood.*

Some people think that the nature and reality of a thing which is due to nature is the primary constituent present in it, ⟨something⟩ unformed in itself. Thus in a bed it would be the wood, in a statue the bronze. It is an indication of this, says Antipho, that if you bury a bed, and the decomposition gets the ability to send up a shoot, what comes up will not be a bed but wood: this seems to show that the disposition of parts customary for beds and the artistry belong only by virtue of concurrence, and that the reality is that which persists uninterruptedly while being affected in these ways. And if the particular kinds of material too are related to something else in the same way, if, for instance, bronze and gold stand thus to water, and bone and wood to earth, and so on, the thing to which they stand in this relation will be their nature and reality. Hence fire, earth, air, and water have been held to be the nature of things, some people choosing just one for this role, some several, and some making use of all. Those who fix on some such element or elements represent it or them as the entire reality, and say that other things are merely affections, states, or dispositions; and these elements are all held to be imperishable in that they do not change out of themselves, whilst other things come to be and pass away as often as you please.

That is one way of using the word 'nature': for the primary underlying matter in each case, of things which have in themselves a source of their movements and changes. It is also used for the shape and form which accords with a thing's account. Just as that which is in accordance with art and artificial is called art, so that which is in accordance with nature and natural is called nature. And as in the one case we would not yet say that a thing is at all in accordance with art, or that it is art, if it is a bed only in possibility, and has not yet the form of a bed, so with things constituted naturally: that which is flesh or bone only in possibility, before it acquires the form which accords with the account by which we define what flesh or bone is, does not yet have its proper nature, and is not a thing due to nature. So there is another way of speaking, according to which nature is the shape and form of things which have in themselves a source of their changes, something which is not separable except in respect of its account. Things which consist of this and the matter together, such as men, are not themselves natures, but are due to nature.

The form has a better claim than the matter to be called nature. For we call a thing something, when it is that thing in actuality, rather than just in possibility.

Further, men come to be from men, but not beds from beds. That is why people say that the nature of a bed is not the shape but the wood, since if it sprouts, what comes to be is wood and not a bed. But if this shows that the wood is nature, nature is form too; for men come to be from men.

Again, nature in the sense in which the word is used for a process proceeds towards nature. It is not like doctoring, which has as its end not the art of medicine but health. Doctoring must proceed from the art of medicine, not towards it. But the process of growth does not stand in this relation to nature: that which is growing, as such, is proceeding from something to something. What, then, is it which is growing? Not the thing it is growing out of, but the thing it is growing into. So the form is nature.

Things may be called form and nature in two ways, for the

lack is a form in a way. But whether or not there is a lack and an opposite involved in cases of simply coming to be, we must consider later.

CHAPTER 2

Having distinguished the various things which are called nature, we must go on to consider how the student of mathematics differs from the student of nature—for natural bodies have planes, solids, lengths, and points, which are the business of the mathematician. And again, is astronomy a branch of the study of nature, or a separate subject? It would be absurd if the student of nature were expected to know what the sun or moon is, but not to know any of the things which of themselves they have supervening on them, especially as it is plain that those who discuss nature do also discuss the shape of the sun and moon, and whether the earth and the cosmos are spherical or not.

Both the student of nature and the mathematician deal with these things; but the mathematician does not consider them as boundaries of natural bodies. Nor does he consider things which supervene as supervening on such bodies. That is why he separates them; for they are separable in thought from change, and it makes no difference; no error results. Those who talk about ideas do not notice that they too are doing this: they separate physical things though they are less separable than the objects of mathematics. That becomes clear if you try to define the objects and the things which supervene in each class. Odd and even, straight and curved, number, line, and shape, can be defined* without change but flesh, bone, and man cannot. They are like snub nose, not like curved. The point is clear also from those branches of mathematics which come nearest to the study of nature, like optics, harmonics, and astronomy. They are in a way the reverse of geometry. Geometry considers natural lines, but not as natural; optics treats of mathematical lines, but considers them not as mathematical but as natural.

Since there are two sorts of thing called nature, form and matter, we should proceed as if we were inquiring what snubness is: we should consider things neither without their matter nor in accordance with their matter. For it is certainly a problem, if there are two sorts of nature, which of them the student of nature is concerned with. Perhaps with that which consists of the two together. In that case he will be concerned with both. Will both, then, fall under the same study, or each under a different? If we had regard to the early thinkers, it might seem that the study of nature is the study of matter, for Empedocles and Democritus touched only very superficially on form and what the being would be. But if art imitates nature, and it belongs to the same branch of knowledge to know the form and to know the matter up to a point (thus the doctor has knowledge of health, and also of bile and phlegm, the things in which health resides; and the builder knows the form of a house, and also the matter—that it is bricks and beams; and it is the same with other arts), then it belongs to the study of nature to know both sorts of nature.

Further, it belongs to the same study to know the end or what something is for, and to know whatever is for that end. Now nature is an end and what something is for. For whenever there is a definite end to a continuous change, that last thing* is also what it is for; whence the comical sally in the play 'He has reached the end for which he was born'—for the end should not be just any last thing, but the best.

Indeed, the arts make their matter, that is, they either bring it into being altogether, or render it good to work with; and we use all things as if they were there for us. (For we too are ends of a sort. As was said in the *De philosophia*, there are two sorts of thing which a thing may be said to be for.) There are two arts which control the matter and involve knowledge, the art of using, and the art which directs the making. Hence the art of using too is directive in a way, but is different in that it involves knowledge of the form, whilst the art which is directive in that it is the art of making involves knowledge of the matter. The steersman knows and prescribes what the form for a rudder is, and the carpenter knows out of what sort of wood and by what changes it will be made. In the case, then, of artefacts we make the matter for the work to be done, whilst in the case of natural objects it is there already.

Again, matter is something relative to something, for the matter varies with the form.

Up to what point, then, should the student of nature know the forms of things and what they are? Perhaps he should be like the doctor and the smith, whose knowledge of sinews and bronze extend only to what they are for; and he should confine himself to things which are separable in form, but which are in matter. For a man owes his birth to another man and to the sun. What it is which is separable, and how things are with it, it is the work of first philosophy to determine.

CHAPTER 3

These distinctions having been drawn, we must see if we can characterize and enumerate the various sorts of cause. For since the aim of our investigation is knowledge, and we think we have knowledge of a thing only when we can answer the question about it 'On account of what?' and that is to grasp the primary cause—it is clear that we must do this over coming to be, passing away, and all natural change; so that, knowing their sources, we may try to bring all particular objects of inquiry back to them.

According to one way of speaking, that out of which as a constituent a thing comes to be is called a cause; for example, the bronze and the silver and their genera would be the causes respectively of a statue and a loving-cup. According to another, the form or model is a cause; this is the account of what the being would be, and its genera—thus the cause of an octave is the ratio of two to one, and more generally number—and the parts which come into the account. Again, there is the primary source of the change or the staying unchanged: for example, the man who has deliberated is a cause, the father is a cause of the child, and in general that which makes something of that which is made,* and that which changes something of that which is changed. And again, a thing may be a cause as the end. That is what something is for, as health might be what a walk is for. On account of what does he walk? We answer 'To keep fit' and think that, in saying that, we have given the cause. And anything which, the change being effected by something else, comes to be on the way to the end, as slimness, purging, drugs, and surgical instruments come to be as means to health: all these are for the end, but differ in that the former are works and the latter tools.

That is a rough enumeration of the things which are called causes. Since many different things are called causes, it follows that many different things can all be causes, and not by virtue of concurrence, of the same thing. Thus the art of statue-making and the bronze are both causes of a statue, and causes of it, not in so far as it is anything else, but as a statue; they are not, however, causes in the same way, but the latter is a cause as matter, and the former as that from which the change proceeds. And sometimes two things are causes each of the other; thus labour is the cause of strength, and strength of labour; not, however, in the same way, but the one is a cause as the end, and the other as source of change. And again, the same thing is the cause of opposites. That which, by being present, is the cause of so and so, is sometimes held responsible by its absence for the opposite; thus the loss of a ship is set down to the absence of the steersman, whose presence would have been the cause of its being saved.

All the causes we have mentioned fall into four especially plain groups. Letters are the cause of syllables, their matter of artefacts, fire and the like of bodies, their parts of wholes, and the hypotheses of the conclusion, as that out of which; and the one lot, the parts and so on, are causes as the underlying thing, whilst the other lot, the whole, the composition, and the form, are causes as what the being would be. The seed, the doctor, the man who has deliberated, and in general the maker, are all things from which the change or staying put has its source. And there are the things which stand to the rest as their end and good; for what the other things are for tends to be best and their end. It may be taken as making no difference whether we call it good or apparently good.

That, then, is how many species of cause there are. There are a good many ways in which something can be a cause, but these too may be brought under comparatively few heads. Many different things are said to be causes, and even among causes of the same species some are prior and some posterior; thus the cause of health is a doctor and a man of skill, the cause of an octave is double and number, and always there are the particulars and the genera which embrace them. And some are causes as concurrent, or as the genera of these; thus the cause of a statue is in one way a sculptor and in another Polyclitus, in that being Polyclitus supervenes on the sculptor. Similarly that which embraces what supervenes; thus a man, or more generally an animal, might be the cause of a statue. And of causes by concurrence too, some are further and some nearer, as when we might call a pale man or a musician the cause of the statue. And both proper causes and causes by virtue of concurrence may be spoken of either as able to cause or as actually causing; thus the cause of the building of a house may be called a builder or a builder who is building.

Similarly with the things to which the causes stand as causes. A thing is said to stand as cause to this statue, or to a statue, or more generally to an image; or to this bronze, or to bronze, or more generally to matter. And the same with things which supervene. And a combination, either of causes or of things to which they stand as causes, may be given, as when we say, not that the cause is Polyclitus or that it is a sculptor, but that it is the sculptor Polyclitus.

All this comes to six things, which may each be spoken of in two ways. There is the particular and the general, the concurrent and the genus to which the concurrent belongs, and these may be given singly or in combination. And any of them may be actual or possible. What difference does that make? Those causes which are particular and actual, are and are not simultaneously with the things of which they are causes. Thus there is a particular man actually doctoring as long as there is a particular man actually being healed, and a particular man actually building as long as there is a particular building actually being built. With causes which are merely possible, the same does not always hold: the builder and the house do not pass away at the same time.

As elsewhere, so here, we should look always for the topmost cause of each thing. Thus a man builds because he is a builder, and a builder builds in accordance with the art of building; the art of building, then, is the prior cause, and similarly in all cases. Again, we should look for kinds of cause for kinds of thing, and particular causes for particular things. Thus a sculptor is the cause of a statue, and this sculptor here of this statue here. And we should look for abilities as causes for things which are possible, and things actually causing for things which are being actualized.

On how many causes there are, and in what ways they are causes, let these distinctions suffice.

CHAPTER 4

Luck and the automatic are reckoned as causes, and we say that many things are and come to be on account of them. We must see, then, in what way luck and the automatic fit into our causes, whether luck and the automatic are the same or different, and in general what they are.

Some people wonder even whether there are any such things
or not. They say that nothing comes to be as an outcome of 196^{a}
luck, but that there is a definite cause of everything which we say comes to be as an automatic outcome or as an outcome of luck. Thus when we say that a man as the outcome of luck came into the market-place, and found there someone he wished but did not expect to find, they claim that the cause

was wishing to go and attend the market. And similarly with other things which are said to be the outcome of luck: it is always possible to find some cause for them other than luck; since if there were such a thing as luck, it would seem to be really very absurd, and one might wonder why it is that none of the sages of the past who discussed the causes connected with coming to be and passing away gave any distinct account of luck; but it seems that they too thought that nothing is the outcome of luck.

Yet this too is amazing. Many things come to be, and many things are, as the outcome of luck or as an automatic outcome; and though not unaware that, as the old saw* which does away with luck says, everything which comes to be can be referred back to some cause, still, all men say that some things are an outcome of luck, and others not. Hence they ought to have made mention of it somehow or other. But they cannot be said even to have equated it with any of the causes they recognized, love or strife or mind or fire or the like. Either way, then, it is absurd, whether they did not think there was any such thing, or did but left it aside. Especially when they sometimes make use of it, as Empedocles does when he says that air is separated out on top not invariably, but as luck will have it. At any rate he says in his Cosmogony: 'Thus chanced it to be running then, oft chancing otherwise.' And he says that the parts of animals mostly came to be as the outcome of luck.

There are others who make the automatic responsible for our own heavens and for all the cosmic systems. They say it was an automatic outcome that the Swirl came to be, and the change which separated out and established the universe in its present arrangement. Yet this itself may excite amazement. On the one hand they hold that animals and plants neither come to be nor are as the outcome of luck, but that their cause is nature or mind or something like that—for it is not as luck will have it, what comes to be from a particular seed, but an olive-tree comes to be from one sort and a man from another—but on the other hand they say that the heavens

and the most divine of the things we see, came to be as an automatic outcome, without there being any such cause as animals and plants have. If that is so, that very thing is worthy of attention, and it would have been well to say some- 196b
thing about it.* For not only is what they say absurd in other respects, but it is still more absurd for them to say it when they see nothing in the heavens coming to be as an automatic outcome, whilst in the things which are supposed not to be the outcome of luck they see many things supervening as the outcome of luck. It might have been expected to be the other way round.

There are also some who think that luck is indeed a cause, but one inscrutable to human thought, because it is divine or supernatural in character. So we must examine the automatic and luck, and see what each is, whether they are the same or different, and what their place is among the causes we have distinguished.

CHAPTER 5

In the first place, then, since we see some things always, and others for the most part, coming to be in the same way, it is plain that luck or its outcome is not called the cause of either of these—of that which is of necessity and always, or of that which is for the most part. But since there are other things which come to be besides these, and all men say that they are the outcome of luck, plainly there is such a thing as luck and the automatic; for we know that things of this sort are the outcome of luck, and that the outcome of luck is things of this sort.

Of things which come to be, some come to be for something, and some do not. Of the former, some are in accordance with choice and some are not, but both are among things which are for something. Clearly, then, also among things which are neither necessary nor for the most part, there are some to which it can belong to be for something. Anything which might be done as an outcome of thought or nature is for something. Whenever something like this comes to be by virtue of concurrence, we say that it is the outcome of luck. For as a thing is, so it can be a cause, either by itself or by virtue of concurrence. Thus that which can build is by itself the cause of a house, but that which is pale or knows music is a cause by virtue of concurrence. That which by itself is a cause is determinate, but that which is a cause by virtue of concurrence is indeterminate; for an unlimited number of things may concur in the one.

As has been said, then, whenever this happens over something which comes to be for something, it is said to be an automatic outcome or the outcome of luck. (The difference between these two we shall have to determine later; for the moment this much is plain, that both are to be found among things which are for something.) Thus the man would have come for the purpose of getting back the money when his debtor was collecting contributions, if he had known; in fact, he did not come for this purpose, but it happened concurrently that he came, and did what was for getting back the money.* And that, though he used to go to the place
197a neither for the most part nor necessarily. The end, the recovery, is not one of the causes in him, but it is an object of choice and an outcome of thought. And in this case the man's coming is said to be the outcome of luck, whilst if he had chosen and come for this purpose, or used to come always or for the most part, it would not be called the outcome of luck. Clearly, then, luck is a cause by virtue of concurrence in connection with those among things for something which are objects of choice. Hence thought and luck have the same field, for choice involves thought.

Necessarily, then, the causes from which an outcome of luck might come to be are indeterminate. That is why luck is thought to be an indeterminate sort of thing and inscrutable to men, and at the same time there is a way in which it might be thought that nothing comes to be as the outcome of luck. For all these things are rightly said, as might be expected. There is a way in which things come to be as the outcome of luck: they come to be by virtue of concurrence, and luck is a concurrent cause. But simply, it is the cause of nothing. As in the case of a house the cause is a builder, but by virtue of concurrence a flute-player, so in the case of the man who came and recovered the money, but did not come for that purpose, an unlimited number of things can be causes by concurrence. He might have been hoping to see someone, or litigating as plaintiff or defendant, or going to the theatre. And it is right to oppose luck to the accountable. We account for that which is always or for the most part, and luck appears in the cases apart from these. So since the causes in such cases are indeterminate, so is luck. Still, there are cases which may raise the doubt: could anything whatsoever come to be a cause of luck? For instance could the breath of the wind or the warmth of the sun be the cause of health, but not having had a haircut? For of things which are causes by virtue of concurrence, some are nearer than others.

Luck is called good when something good comes out, and bad when something bad, and it is called good fortune or bad fortune when the consequences are sizable. Hence just to miss meeting with a great evil or good is to be lucky or unlucky, for thought treats the good or evil as already yours; what is so close seems no distance off at all. That good fortune is inconstant is also to be expected; for luck is inconstant; nothing which is the outcome of luck can be either always or for the most part.

As has been said, then, luck and the automatic are both causes by virtue of concurrence, in the field of things which are capable of coming to be neither simply* nor for the most part, and of such of these as might come to be for something.

CHAPTER 6

They differ in that the automatic extends more widely. Everything which is the outcome of luck is an automatic outcome,

197b but not everything which is the latter is the outcome of luck. For luck and its outcome belong only to things which can be lucky and in general engage in rational activity. Hence luck must be concerned with things achievable by such activity. It is an indication of this that good fortune is thought to be the same as happiness or close to it, and happiness is a kind of rational activity: it is activity going well. So what is incapable of such activity, can do nothing as the outcome of luck.

Hence nothing done by an inanimate object, beast, or child, is the outcome of luck, since such things are not capable of choosing. Nor do good or bad fortune belong to them, unless by a resemblance, as Protarchus said that lucky are the stones from which altars are made, since they are honoured, whilst their fellows are trodden underfoot. In a way these things can undergo something as the outcome of luck, when a person engaged in activity concerning them achieves something as an outcome of luck; but otherwise not.

The automatic, on the other hand, extends to the animals other than man and to many inanimate objects. Thus we say that the horse came automatically, in that it was saved because it came, but it did not come for the purpose of being saved. And the tripod fell automatically. It was set up for someone to sit on, but it did not fall for someone to sit on. Plainly, then, in the field of things which in a general way come to be for something, if something comes to be but not for that which supervenes, and has an external cause, we say that it is an automatic outcome; and if such an outcome is for something capable of choosing and is an object of choice, we call it the outcome of luck.

An indication is the expression 'in vain', which we use when something is for something else, and what it is for does not come to be.* For instance, suppose walking is for the loosening of the bowels, and a man walks without having this come to be: we say that he walked in vain and that his walk was vain, suggesting that this is what is in vain: something which is by nature such as to be for something else, when it does not accomplish that which it was for and which it is by nature such as to be for—since if someone said that he had performed his ablutions in vain because the sun did not go into eclipse, he would be ridiculous. Solar eclipses are not what washing is for. This, then, is what the automatic is like when it comes to be in vain, as the word itself suggests. The stone did not fall for the purpose of hitting someone; it fell, then, as an automatic outcome, in that it might have fallen through someone's agency and for hitting.

We are furthest from an outcome of luck with things which come to be due to nature. For if something comes to be contrary to nature, we then say not that it is the outcome of luck but rather that it is an automatic outcome. Yet it is not quite that either: the source of an automatic outcome is external, whilst here it is internal.

What the automatic and luck are, then, and how they differ, has now been said. As for the ways in which they are causes, both are sources from which the change originates; for they are always either things which cause naturally or things which cause from thought—of which there is an indeterminate multitude. But since the automatic and luck are causes of things for which mind or nature might be responsible, when something comes to be responsible for these same things by virtue of concurrence,* and since nothing which is by virtue of concurrence is prior to that which is by itself, it is clear that no cause by virtue of concurrence is prior to that which is by itself a cause. Hence the automatic and luck are posterior to both mind and nature; so however much the automatic may be the cause of the heavens, mind and nature are necessarily prior causes both of many other things and of this universe.

CHAPTER 7

That there are causes, and that they are as many as we say, is clear: for that is how many things the question 'On account of what?' embraces. Either we bring it back at last to the question 'What is it?'—that happens over unchangeable things; for instance in mathematics it comes back at last to a definition of straight or commensurable or the like. Or to that which in the first instance effects the change; thus on account of what did they go to war? Because of border raids. Or it is what the thing is for: they fought for dominion. Or, in the case of things which come to be, the matter.

Plainly, then, these are the causes, and this is how many they are. They are four, and the student of nature should know about them all, and it will be his method, when stating on account of what, to get back to them all: the matter, the form, the thing which effects the change, and what the thing is for. The last three often coincide. What a thing is, and what it is for, are one and the same, and that from which the change originates is the same in form as these. Thus a man gives birth to a man, and so it is in general with things which are themselves changed in changing other things—and things which are not so changed fall beyond the study of nature. They have no change or source of change in themselves when they change other things, but are unchangeable. Hence there are three separate studies: one of things which are unchangeable, one of things which are changed but cannot pass away, and one of things which can pass away.

So in answering the question 'On account of what'? we bring it back to the matter, and to what the thing is, and to what first effected the change. People usually investigate the causes of coming to be thus: they see what comes after what, and what first acted or was acted on, and go on seeking what comes next. But there are two sources of natural change, of which one is not natural, since it has no source of change in itself. Anything which changes something else without itself being changed is of this latter sort; for instance, that which is completely unchangeable and the first thing of all, and a thing's form or what it is, for that is its end and what it is for. Since, then, nature is for something, this cause too should be known, and we should state on account of what in every way: that this out of this necessarily (i.e. out of this simply, or out of this for the most part); and if so and so is to be (as the conclusion out of the premisses); and that this would be what the being would be; and because better thus—better not simply, but in relation to the reality of the thing concerned.

CHAPTER 8

We must first give reasons for including nature among causes which are for something, and then turn to the necessary, and see how it is present in that which is natural. For everyone brings things back to this cause, saying that because the hot is by nature such as to be thus, and similarly the cold and everything else of that sort, therefore these things of necessity come to be and are. For if they mention any other cause, as one does love and strife and another mind, they just touch on it and then goodbye.

The problem thus arises: why should we suppose that nature acts for something and because it is better? Why should not everything be like the rain? Zeus does not send the rain in order to make the corn grow: it comes of necessity. The stuff which has been drawn up is bound to cool, and having cooled, turn to water and come down. It is merely concurrent that, this having happened,* the corn grows. Similarly, if someone's corn rots on the threshing-floor, it does not rain for this purpose, that the corn may rot, but that came about concurrently. What, then, is to stop parts in nature too from being like this—the front teeth of necessity growing sharp and suitable for biting, and the back teeth broad and serviceable for chewing the food, not coming to be *for* this, but by coincidence? And similarly with the other parts in which the 'for something' seems to be present. So when all turned out just as if they had come to be for something, then the things, suitably constituted as an automatic outcome, survived; when not, they died, and die, as Empedocles says of the man-headed calves.

This, or something like it, is the account which might give us pause. It is impossible, however, that this should be how things are. The things mentioned, and all things which are

due to nature, come to be as they do always or for the most part, and nothing which is the outcome of luck or an auto-
199a matic outcome does that. We do not think that it is the outcome of luck or coincidence that there is a lot of rain in winter, but only if there is a lot of rain in August; nor that there are heatwaves in August, but only if there is a heatwave in winter. If, then, things seem to be either a coincidental outcome or for something, and the things we are discussing cannot be either a coincidental or an automatic outcome, they must be for something. But all such things are due to nature, as the authors of the view under discussion themselves admit. The 'for something', then, is present in things which are and come to be due to nature.

Again, where there is an end, the successive things which go before are done for it. As things are done, so they are by nature such as to be, and as they are by nature such as to be, so they are done, if there is no impediment. Things are done for something. Therefore they are by nature such as to be for something. Thus if a house were one of the things which come to be due to nature, it would come to be just as it now does by the agency of art; and if things which are due to nature came to be not only due to nature but also due to art, they would come to be just as they are by nature. The one, then, is for the other. In general, art either imitates the works of nature or completes that which nature is unable to bring to completion. If, then, that which is in accordance with art is for something, clearly so is that which is in accordance with nature. The relation of that which comes after to that which goes before is the same in both.

The point is most obvious if you look at those animals other than men, which make things not by art, and without carrying out inquiries or deliberation. Spiders, ants, and the like have led people to wonder how they accomplish what they do, if not by mind. Descend a little further, and you will find things coming to be which conduce to an end even in plants, for instance leaves for the protection of fruit. If, then, the swallow's act in making its nest is both due to nature and for something, and the spider's in making its web, and the plant's in producing leaves for its fruit, and roots not up but down for nourishment, plainly this sort of cause is present in things which are and come to be due to nature. And since nature is twofold, nature as matter and nature as form, and the latter is an end, and everything else is for the end, the cause as that for which must be the latter.*

Mistakes occur even in that which is in accordance with art. Men who possess the art of writing have written incorrectly, doctors have administered the wrong medicine. So clearly the same is possible also in that which is in accordance with nature. If it sometimes happens over things which are in accordance with art, that that which goes right is for something, and that which goes wrong is attempted for something but miscarries, it may be the same with things which are natural, and monsters may be boss shots at that which is for something. When things were originally being constituted, man-headed calves, if they were unable to reach a certain limit and end, came to be as a result of a defect in some principle, as they now do as the result of defective seed.

Again, seed must come first, and not the animal straight off, and the 'omnigenous protoplast'* was seed.

Again, the 'for something' is present in plants too, though it is less articulate. Was it the case, then, that as there were man-headed calves, so there were olive-headed vinelets in the vegetable kingdom? Or is that absurd? But there should have been, if that is how it was with animals.

Again, coming to be among seeds too would have had to be as luck would have it. But a person who says that does away with nature and things due to it altogether. A thing is due to nature, if it arrives, by a continuous process of change, starting from some principle in itself, at some end. Each principle gives rise, not to the same thing in all cases, nor to any chance thing, but always to something proceeding towards the same thing, if there is no impediment. What something is for, and what is for that, can also come to be as the outcome of luck, as when we say that the family friend came as the outcome of luck and paid the ransom before departing, if he behaved as if he had come for that purpose but had not in fact come for that purpose. That is by virtue of concurrence (for luck is a cause by virtue of concurrence, as we said above); but when a certain thing comes to be always or for the most part, it is not a concurrent happening, nor the outcome of luck. Now with that which is natural it is always thus if there is no impediment.

It is absurd not to think that a thing comes to be for something unless the thing which effects the change is seen to have deliberated. Art too does not deliberate. If the art of shipbuilding were present in wood, it would act in the same way as nature; so if the 'for something' is present in art, it is present in nature too. The point is clearest when someone doctors himself: nature is like that.

That nature is a cause, then, and a cause in this way, for something, is plain.

CHAPTER 9

Is that which is of necessity, of necessity only on some hypothesis, or can it also be simply of necessity? The general view is that things come to be of necessity, in the way in which a man might think that a city wall came to be of necessity, if he thought that since heavy things are by nature such as to sink down, and light to rise to the surface, the stones and foundations go down, the earth goes above them because it is lighter, and the posts go on top because they are lightest of all. Now without these things no city wall would have come to be; still, it was not on account of them, except as matter, that it came to be, but for the protection and preservation of certain things. Similarly with anything else in which the 'for something' is present: without things which have a necessary nature it could not be, but it is, not on account of them, except in the way in which a thing is on account of its matter, but for something. Thus on account of what is a saw like this?

That this may be, and for this. It is impossible, however, that this thing which it is for should come to be, unless it is made of iron. It is necessary, then, that it should be made of iron, if there is to be a saw, and its work is to be done. The necessary, then, is necessary on some hypothesis, and not as an end; the necessary is in the matter, the 'that for which' in the account.

The necessary appears in mathematics and in the things which come to be in accordance with nature, in a parallel fashion. Because the straight is so and so, it is necessary that a triangle should have angles together equal to two right angles, and not the other way round. Still, if triangles did not have angles together equal to two right angles, we should have no straight lines.* With things which come to be for something the case is reversed: if the end will be or is, that which comes before will be or is; and if we do not have it, then just as in mathematics, if we do not have the conclusion, we shall not have the starting-point, so here we shall not have the end or that for which. That too is a starting-point, not of the practical activity, but of the reasoning. (In mathematics too the starting-point is of the reasoning, since there is no practical activity there.) So if there is to be a house, it is necessary that these things should come to be or be present, and in general it is necessary that there should be the matter which is for something, e.g. the bricks and stones if there is to be a house. Nevertheless, the end is not on account of these things except as matter, nor on account of them will it come into being. In general, if they, for instance the stones or the iron, are not present, there will be no house or saw; just as in mathematics there will not be the starting-points if the triangle does not have angles together equal to two right angles.

Plainly, then, the necessary in things which are natural is that which is given as the matter, and the changes it undergoes. The student of nature should state both causes, but particularly the cause which is what the thing is for; for that is responsible for the matter, whilst the matter is not responsible

for the end. And the end is that for which, and the start is from the definition and the account; and just as in the case of things which are in accordance with art, since this is the
200b sort of thing a house is, this and that must of necessity come to be and be present, or since this is what health is, this and that must come to be of necessity and be present: so if this is what a man is, then so and so, and if so and so, then such and such.

Perhaps the necessary enters even into the account. Suppose the work of sawing is defined as a certain sort of division: that will not be, unless the saw has teeth of a certain sort, and there will not be teeth like that, if it is not made of iron. For even in the account there are parts which stand to it as matter.

NOTES ON THE TEXT AND TRANSLATION

185ᵃ25: *λευκός*: this word is commonly translated 'white'; whilst, however, it is used for what is white, it is in fact (from our point of view; the Greeks seem to have classified colours differently) a vaguer word approximating to our 'pale', and I have translated it consistently 'pale', and *μέλας* (in I. 5 and elsewhere) 'dark'. Pale and dark are indefinite opposites, like hot and cold, large and small, whilst white and black are definite colours, as blood-heat, bath-temperature are definite temperatures; and indefinite opposites often seem to be what Aristotle has in mind when he speaks of *λευκός* and *μέλας*, e.g. 189ᵃ18–19, cf. also Plato, *Theaet.* 154 b 1–2, 182 a 3. And Aristotle speaks of musicians (188ᵃ35 ff.) and doctors (191ᵇ5 ff.) changing from *λευκός* to *μέλας* and back; he is thinking, not of nigger minstrels or miraculous solutions to the colour problem, but of the change from being pale to being sunburnt or vice versa. (Such changes will have been more striking in Greece than in the British Isles, and it was a matter for comment whether a man was pale or sunburnt, cf. Euripides, *Bacchae* 457–8, Xenophon, *Hellenica* III. iv. 19.)

185ᵇ33: *μουσικός*: I translate this word 'knowing music' or (as here) 'musician', but it was often used in a wider sense, and Aristotle may mean by it what we mean by 'cultured' or 'polished'.

186ᵃ7–10: *καὶ γὰρ . . . χαλεπόν*: these lines are bracketed by Ross because of their resemblance to 185ᵃ9–12. If, however, chapters 2 and 3 were originally alternative lectures (v. infra, p. 53) the same gibe may have appeared in both.

186ᵇ15–16: reading *εἰ ἔστιν ὅπερ ὄν*, with E and, perhaps, Philoponus. Ross reads *ὅπερ ὄν τι* 'suppose a man is a thing which is precisely what is', which would support his interpretation against that preferred in the commentary.

186b20–1: either ἢ οὗ ἐν . . . συμβέβηκεν or ἢ ἐν ᾧ . . . συμβέβηκεν should no doubt be bracketed. Ross brackets the latter, which is inferior as a formulation of what Aristotle means, but the wording of the example b22–3 suggests that it should stay and the former go.

187a5–6: reading μὴ ἁπλῶς μὴ εἶναι with F and, perhaps, Alexander. Ross reads μὴ ἁπλῶς εἶναι and translates: 'for there is nothing to prevent that which is not—not from being, simply, but from being what is not some particular thing'.

187a26: Ross takes καλούμενα as ironical: the *so-called* elements (though of course they are not really elements at all). In fact, στοιχεῖον strictly means a letter or phoneme, and was perhaps first applied to earth, air, fire, and water only by Plato; I think, therefore, that τὰ καλούμενα στοιχεῖα means: 'the things philosophers call the elements'.

188a27: μήτε ἐξ ἀλλήλων εἶναι: 'the principles must come' etc., or 'the principles must consist, neither of one another' etc. Aristotle probably means both.

188b12–14: ἡρμοσμένον . . . ἀνάρμοστον: 'united . . . disunited', or perhaps, as Ross and others, 'tuned . . . untuned', referring to the strings of a lyre.

189a18: reading ἐξ ἄλλων with EVS. If, like Ross, we read ἐξ ἀλλήλων with FI, we must translate 'from one another', and it is unlikely that Aristotle thought that sweet and sour arise from pale and dark and vice versa.

190a11: μουσικὸς γιγνόμενος ἄνθρωπος καὶ ἔστι: 'and is (sc. still) a man' etc., or perhaps: 'when he comes to be a knowing-music man, and still exists'.

190b18–19: μὴ κατὰ συμβεβηκὸς ἀλλ' ἕκαστον ὃ λέγεται κατὰ τὴν οὐσίαν: Ross paraphrases: 'Evidently, then, if the elements from which natural things are and have come to be—not possessed of some accident, but what they essentially are—' etc., suggesting that Aristotle is here speaking of coming into existence in contrast with alteration. However, Aristotle immediately (b20–2) goes on to illustrate his point with a case of alteration; his argument developed from considerations about alterations (189a34 ff.); he claims to be giving an account of what is common to all cases of becoming (189b30–1); and he does not in the rest of the chapter (see 190b28–32,

191^{a}1–3) or indeed of the book, seem to distinguish alterations from comings into existence. I think, therefore, that the contrast he has in mind, here as below, 190^{b}25–7, 191^{b}14–15, is between principles *καθ' αὑτό* like the underlying thing, and principles *κατὰ συμβεβηκός* like the lack. On this view, *μὴ κατὰ συμβεβηκός* and *κατὰ τὴν οὐσίαν* go rather with *ἐξ ὧν* than with *γεγόνασι*, and Aristotle either means by *κατὰ τὴν οὐσίαν* 'strictly', 'in truth' (cf. *Met. Δ* 1019^{a}3), or else (as is suggested in my translation and by 190^{b}22–3) has in mind the point (194^{a}16–17 etc.) that the account of the *οὐσία* of a thing should specify both matter and form. A saw is by definition 'out of' iron, a man 'out of' flesh and bone (200^{b}5–8, *Met. Z* 1034^{a}6). For *οὐσία* in connection with things other than realities in the strict sense like animals and plants cf. *De an.* II 418^{a}25.

191^{a}3: I bracket *περὶ γένεσιν*, as Ross suggests.

191^{a}17–18: *τίς ἡ διαφορὰ τῶν ἐναντίων*: Ross paraphrases: 'we have now shown the difference between the contraries'; similarly Hardie and Gaye. This, however, is extremely weak, and the phrase literally means: 'what the differentiating feature of the opposites is', i.e. what differentiates them from other pairs of opposites, like the hot and the cold. For Aristotle's use of *διαφορά* as 'differentiating feature', v. *Met. H* 1042^{b}12, 15, 31. If he had wanted to say 'we have shown the difference between the opposites', he could have written: *πῶς διαφέρει τὰ ἐναντία*.

191^{b}36–192^{a}1: *ὁμολογοῦσιν ἁπλῶς γίγνεσθαί τι ἐκ μὴ ὄντος*. I take *ἁπλῶς* with *ὁμολογοῦσιν*: 'they agree in a general, uncritical way, without making the proper distinctions', cf. *De an.* III 426^{a}26, *E.E.* I 1218^{a}27. Even if we take it with *γίγνεσθαι*, we should probably still understand: 'they agree that it is on the whole true that things come to be . . .', cf. 191^{b}14, 197^{a}14, ^{b}19, rather than 'they agree that things *come into existence*'—*γίγνεσθαι ἁπλῶς* as contrasted with *γίγνεσθαί τι*. Aristotle does not here seem concerned with the distinction between coming into existence and coming to be something, and elsewhere he says that Plato's conception of the underlying thing leaves no room for it (*De gen. et cor.* II 329^{a}13–21). Cherniss (pp. 92–3) says that Aristotle misrepresents Plato as holding that things come to be out of absolute non-being. If Aristotle were doing that here, he would have written *ἐξ ἁπλῶς μὴ ὄντος* or *ἐκ μὴ ὄντος ἁπλῶς*.

192a26–7: *τὸ γὰρ . . . στέρησις*: bracketed by myself; see commentary, p. 83.

193a6–9: I follow Ross in taking *ὅτι δ' ἐνδέχεται . . . χρωμάτων* as parenthetical, and *τοῖς τοιούτοις*, accordingly, as referring not to the man blind from birth, who in making inferences about colours is doing the best he can, but to the man who in philosophy cannot distinguish what is self-evident from what is not. If we remove Ross's brackets, *οἱ τοιοῦτοι* will be primarily at least men blind from birth: they can argue about words for colours, but cannot know what those words mean. (Cf. Burke, *The Sublime and the Beautiful*, v. 5.)

194a3: I take *ἔσται* as 'it will be possible' sc. to define them. It could be taken to mean 'they will exist'; but it is doubtful if Aristotle thought that straight, curved, etc., could thus exist separately: v. *De an.* I 403a12–16, *Met. M.* 3.

194a30: reading *τοῦτο ἔσχατον* with the MSS. If with Ross we read *τοῦτο ⟨τὸ⟩ ἔσχατον*, the meaning would be: 'this is the last thing and what it is for', which seems weaker.

194b31: *τὸ ποιοῦν τοῦ ποιουμένου*: 'that which makes something of that which is made' or 'the doer of that which is done', *τὸ ποιοῦν* in the sense in which it is contrasted with *τὸ πάσχον*. Aristotle probably means both.

196a14: *ὁ παλαιὸς λόγος*: 'the old saw' (cf. Plato, *Laws* V 715 e 8), or perhaps, as Ross says, 'the before-mentioned argument', given in a1–7.

196a36–b1: *καλῶς ἔχει λεχθῆναί τι περὶ αὐτοῦ*: this might be taken as ironical: 'It is well they mentioned it', but the Greek commentators take it as I have translated it.

196b35–6: perhaps we should read *τοῦτο ⟨τὸ⟩ τοῦ κομίσασθαι ἕνεκα* for *τοῦτο τοῦ κομίσασθαι ἕνεκα*. Ross retains the MSS. reading, and comments: 'We have found that in b21 *ἕνεκά του* means "producing an end-like result" . . . and *ἕνεκα* can have the same meaning here'. I find this obscure; perhaps Ross would wish to translate: 'he came, and did so with the end-like result that he got back the money'; but that seems far-fetched. Others omit *τοῦ κομίσασθαι ἕνεκα*, and understand: 'He did not come for getting back the money, but it happened concurrently that he came and did get back the money'. I translate my suggested reading: 'He came and did this thing (sc. going to where his debtor was) which was for (i.e. which might have been done through thought for) getting back the money'. For the awkward phrase *τοῦτο τὸ τοῦ κ. ἕνεκα* cf. 199ᵇ4.

197ᵃ34: *ἁπλῶς*: 'simply' here equivalent to *ἀεί*, 'always'.

197ᵇ23: reading *μὴ γένηται τῷ ἕνεκα ἄλλου ἐκεῖνο οὗ ἕνεκα*, with Prantl. Ross reads *μὴ γένηται τὸ ἕνεκα ἄλλου ἐκείνου ἕνεκα*, and translates: 'when that which is intended to produce a result other than itself does not produce it.' I find this sense hard to extract from his text.

198ᵃ6–7: *ὅταν . . . αὐτῶν*: it would be possible to take *τούτων αὐτῶν* as a partitive genitive after *τι*, referring to *νοῦς ἢ φύσις*, and understand 'when mind or nature comes to be a cause by virtue of concurrence'; but it is more natural to take the phrase as objective genitive after *αἴτιον*, referring to *ὧν* in ᵃ6.

198ᵇ20–1: *τούτου γενομένου*: this could be taken as causal: 'because this happens'. The point is not that there is no causal connection between the falling of the rain and the growing of the corn, but that the rain does not fall with the purpose of making the corn grow.

199ᵃ30–2: *καὶ ἐπεὶ . . . ἡ οὗ ἕνεκα*: it would not be impossible, though in the absence of an *ἔτι* or the like it is awkward, to take Aristotle as here offering a fresh argument for taking nature as a cause *ἕνεκά του*: 'And again, since nature is not only matter but also form, and form is the end, and the end is what everything else is for, it follows that this cause, namely that for which, is the cause'. If this is indeed Aristotle's argument, then Aristotle is assuming, what he should surely be trying to prove, that the cause of natural things is nature in the sense of form. My translation reflects the view of the Greek commentators (v. Simplicius ad loc.) that Aristotle, having completed (at least for the moment) his argument that nature is a cause *ἕνεκά του*, is now pointing out the consequence that nature is form rather than matter. Ross prints a comma after *αἰτία* which I think is better omitted.

199^b9: *οὐλοφυὲς μὲν πρῶτα*: the whole line (Diels–Kranz 31 B 62, 4) is:

οὐλοφυεῖς μὲν πρῶτα τύποι χθονὸς ἐξανέτελλον

('At first undifferentiated shapes of earth arose', Dr. K. Freeman, *Ancilla to the pre-Socratic philosophers*).

200^a19: *οὐδὲ τὸ εὐθὺ ἔστιν*: 'we should have no straight lines' or, perhaps, 'a straight line would not be what we said'.

COMMENTARY

BOOK I

CHAPTERS 1–2 (184^{a}10–^{b}25)

THE main theme of this introductory section is that the systematic study of nature must start with the attempt to be clear about principles. Some points of terminology:

'principle', *archē*: the central meaning of this word is 'beginning'; in Book I I normally translate it 'principle', and in Book II 'source'.

'cause', *aition*: see on *Phys.* II. 3.

'element', *stoicheion*: see textual note on 187^{a}26. 'Element', 'cause', and 'principle' are here used almost as synonyms, but for their difference in nuance, v. *Met.* *Λ* 4.

'systematic knowledge', *epistēmē*: this word can be restricted to knowledge of things which can be proved, like the propositions of geometry (in which case it means dispositional knowledge of the proof: *An. po.* II 90^{b}9–10). It is also used, however, of disciplines which do not make use of strict proofs: v. *Soph. elench.* 172^{a}28, cf. 12–13.

'nature', *physis*: it appears from *Phys.* II that Aristotle does not recognize any such thing as nature over and above the natures of particular things, but here the word is used for physical things generally.

The natural course, says Aristotle, **184^{a}16–21**, is to start with what is clear to us, and move on to what is clear 'by nature'. Similar remarks are common in Aristotle, e.g. *E.N.* I 1095^{a}2–4, *De an.* II 413^{a}11–12, *Met.* *Z* 1029^{b}3–12. It is not clear, however, that Aristotle always has the same point in mind. In *E.E.* I 1216^{b}26–35 (and cf. 1217^{a}19–2[illegible]) [illegible] speaks of passing from things which are said truly but unclearly to things which are said clearly: the former are the ordinary man's expressions of his intuitions, and are clear to him; the latter are philosophical formulations, and are clear in themselves. (Thus that happiness is the best thing for us, is clear to us; that it is doing what is distinctive of men in the best possible way, is, Aristotle thinks, clear in itself.) Now are the things clear by nature or to us in the other passages formulations, or are they *entities*? In the *De an.* and *E.N.* passages they seem to be formulations, but in the *Met.* *Z* passage and here Aristotle might seem to be thinking of entities. The word 'compounded' in our l. 22 does not support this interpretation, since the same word is used in the *E.E.* passage of opaque or befuddled formulations; but the elements or principles of a23 are apparently the things clear by nature, and they are not formulations. Similarly the things clear to us in *Met. Z* seem to be not formulations but perceptible objects.

I do not think, however, that either passage warrants the view that Aristotle held that there are realms of more and less intelligible entities. In our passage here, we may take him as saying: 'of the ultimate constituents of matter, pure water, pure fire, or whatever they may be, we know little; about things like houses and doctors, on the other hand, we are fairly clear; let us begin, then, with them, and see what emerges, from a discussion of such familiar objects, about basic principles'. And in the *Met. Z* passage his thought is similar.

184a23–b14 are obscure, particularly as we are told elsewhere that the individual is known before the universal (*An. po.* II. 19), and that what we perceive is individual (e.g. *E.N.* VII 1147a26). The general point is probably that made at the beginning of chapter 7: 'The natural course is first to give a general account and then consider the peculiarities of particular cases.' In *Phys.* I Aristotle will talk about the principles of physical objects generally, without distinguishing between products of nature like plants and animals, and products of art like statues and houses. (In a way this is reasonable: having set up his general form–matter distinction, Aristotle can later inquire into the formal and material elements in different sorts of thing. On the other hand, his distinctions are originally set up only through a consideration of particular cases: see especially 188a31–b23).

The point about perception, **a24–5**, might be concerned with parts and wholes (we perceive a man more easily than an individual eyelash of a man) but might be concerned with individuals and particulars. The next words 'and a universal is a sort of whole' follow more naturally if we accept the first interpretation, but the second gets some support from the cryptic remark 'We perceive individuals, but our perception is of the universal, e.g. of man' (or perhaps 'of a man') 'not of the man Callias' (*An. po.* II 100a16–b1: cryptic, because *De an.* II. 6 suggests that forms like man are not strictly objects of perception).

'Words' are said to 'stand in a similar relation to accounts', (**b1**), probably not because definitions make clear the various senses of ambiguous or equivocal expressions (Ross), but because a word like 'man' indicates implicitly a number of features—animal, rational, mortal—which appear separately in the definition (Philoponus).

The point which is meant to emerge from **184b15–25** is that Aristotle's predecessors have all been trying to discover 'the primary constituents of things' (b23), more literally, 'the primary things out of which things are'. Aristotle is not so much accusing them of confining themselves to a search for *material* constituents, as insisting that cosmological inquiry is inquiry after constituents of *some* sort: the question 'What is there?' *ti to on?* can be refined to the question 'What do things ultimately come from?' *ek tinōn prōtōn esti ta onta?* (cf. $190^{b}18$).

CHAPTERS 2–3 ($184^{b}25$–$187^{a}11$)

D. E. Gershenston and D. A. Greenberg have argued (*Phronesis* 1962) that this section consists of two separate arguments against Eleatic monism. The first, they say, running from $184^{b}25$ to $186^{a}32$, is conducted from Aristotle's own point of view, in Aristotelian terminology, and in it various Aristotelian doctrines, notably the doctrine that different things are said to be real for different reasons (which will be explained below), are taken for granted. The second, $186^{a}32$–$187^{a}11$, is an attempt to meet the Eleatics on their own ground, and in it Aristotle argues as far as possible from Eleatic premisses and in Eleatic terminology. In support of this they point out, not only the stylistic difference between the two passages, but also a number of doublets: $185^{b}9$–11 and $186^{b}12$–14, $185^{b}19$–25 and $186^{b}4$–12, $185^{b}16$–19 and $186^{b}14$–35, etc. That we have two separate criticisms of monism seems true, but not that the first runs down to $186^{a}32$. The criticism of Parmenides, $186^{a}22$–32, must, I think, be taken with what follows: it explains why the monist thesis has to be reformulated. If so, the criticism of Melissus, $186^{a}4$–22, should also go with what follows. I take the whole of chapter 3, then, as the alternative argument. The historical note, $185^{b}25$–$186^{a}3$, does well as an end to the first argument, and the most glaring of the doublets, $185^{a}9$–12 and $186^{a}7$–10, supports this division.

In **$184^{b}25$–$185^{a}20$,** Aristotle says that the monist thesis is not a thesis about nature ($^{a}18$) or one to be considered in an inquiry about nature or principles ($^{b}26$, $^{a}3$), but is suitable for discussion by the philosopher ($^{a}20$) or dialectician (the student of what is common, $^{a}2$–3: see above, p. xi). The charge that the monist thesis is not about nature is the more pointed in that Parmenides and Melissus entitled their works 'about nature': can Aristotle substantiate it? He offers, I think, three considerations.

First, the thesis that what is or exists is one and unchangeable does away with principles: **$185^{a}1$–5**. By a principle Aristotle here seems to mean a primary or basic thing: a primary thing must be primary in relation to something else ($^{a}4$–5), and rigid monism forbids a division of what exists into what is primary and what is derivative.

Second, **$^{a}5$–12**, the thesis is wildly paradoxical, and suitable only for dialectical practice, like Heraclitus' thesis that opposite properties belong to everything at the same time (v. *Top.* VIII 159b30–3, 185b20; for Heraclitus' own words v. DK 22 B 59–62, 67, 88, etc.), or like the thesis that reality consists of only one man (perhaps a solipsistic extension of Protagorean doubt).

Third, **a12–17,** it is a basic assumption (for the necessity of making such assumptions cf. *An. po.* I 71a12–16, and for brisk use of them in philosophic argument, *E.E.* II 1218b38, 1219b28) that things are subject to change; and we need deal only with problems which are derived from the appropriate assumptions, just as it is the business of a geometer to refute a quadrature of the circle by means of lunes, but not one like Antipho's. De Morgan in his *Budget of Paradoxes* (p. 389) tells of a man who tried to square the circle by making a circular disk on the lathe, and measuring the diameter and the circumference. It is clearly no business of the geometer to refute a quadrature like that, and Aristotle might be saying that the arguments of Parmenides and Melissus are equally irrelevant. However, it is not clear that they are, and Antipho's quadrature was not like that of de Morgan's paradoxeur. His method (further on the mathematics see Ross ad loc.) was to inscribe polygons with an ever-increasing number of sides, which is the method employed by Euclid, and leads to the value π. His error, if he made one, was that of thinking that if you increase the number of sides without limit, the circumference of the polygon will eventually coincide with the circumference of the circle. This is to do away, as the circle-squarer by means of lunes does not do away, with the fundamental principle that magnitudes are infinitely divisible (Simplicius on the authority of Eudemus). If, then, we take the illustration seriously, Aristotle's charge may be that Parmenides and Melissus do away with some analogous fundamental principle, for instance the principle that natural things are subject to change.

In **185a20–186a3** Aristotle sets out the case against monism generally. Our opening move, he says, should be to point out that 'things are said to be in many ways' (**a21**), literally, that 'being' (the participle of the verb 'to be') 'is said in many ways'. Aristotle is extremely fond of saying that things like being, nature, cause, are said in a number of ways, and an equivalent English formula is hard to find. To say, e.g., 'The word "cause" is used in many senses' is misleading, in that it suggests that Aristotle is talking about a word, when he is in fact talking about the things to which a word is applied. To say 'Causes are spoken of in many ways' is worse, since Aristotle's point is not that many different expressions are applied to the same thing, but that the same expression is applied to many different things. I have used translations varying from, as here, 'things are said to be in many ways', i.e. 'there are many different grounds on which a thing may be said to be a thing which is', to 'many different things can all be called causes' (195a4).

Aristotle recognizes four main 'ways in which a thing may be said', i.e. types of ground on which the same expression may be applied to different things, and since all are relevant to the discussions of *Phys.* I–II, it may be convenient to list them here. First, different things may all be called something on the same grounds, 'in accordance with one thing' (*Met. Z* 1030^b3) or 'in accordance with one idea' (*E.N.* I 1096^a30, $^b25$–6). Thus, perhaps, all the things which are called spherical are called spherical for the same reason, that all points on their surface are equidistant from a single point: we have a single idea of what it is to be spherical, and they are called spherical in so far as they accord with this. Second, things may be called something because they exceed or fall short of some norm (*Phys.* III 200^b29); thus things are called large because they exceed, and small because they fall short of, some norm in size, and there are, of course, various norms relative to which a thing may be called one or the other. All pairs of indefinite opposites, like pale and dark, hot and cold, high and low, seem to be of this sort. Third, things may be called something by analogy (*E.N.* I 1096^b28, cf. *Met. Θ* 1048^a37). In *Met. H.* 2 Aristotle observes that ice is said to exist when water is solidified in a certain way, a threshold when stones are positioned in a certain way, punch when wine and honey are mixed in a certain way, etc. (For criticism, not, I think, fatal, of this interpretation see Owen in Bambrough, *New Essays on Plato and Aristotle*, pp. 78–81.) Since for water to be solidified is not the same as for stones to be arranged, but the one stands to ice as the other to a threshold, we might say that existence is 'said' of things like ice and thresholds by analogy; though Aristotle himself prefers to use the word when the analogy is more far-flung: cf. *Met. Θ* 1048^b6–9, *Λ* 1071^a26–7. Aristotle says that the causes of different things are different but analogous (*Met. Λ* 1070^a31–3), and the class of things which a thing may be called by analogy is probably co-extensive with the class of things it may be called 'according as it acts or undergoes' (*Phys.* III 200^b29–31) (and would include, we may add, many things said in accordance with a family resemblance in Wittgenstein's sense, like 'guide', 'derive'). Finally, things may be called something on the ground that they 'are related to a single thing' (*Met. Γ* 1003^a33–4) in various ways. Thus pink cheeks, mountain-walking, and my brother Henry may all be called healthy, because they are, respectively, indicative of, preservative of, and possessed of, a certain bodily condition. For the example see *Met. Γ* 1003^a34–$^b1$, cf. *Z* 1030^a35–$^b3$, and for an examination of the idea, Owen in Düring and Owen, *Plato and Aristotle in the Mid-Fourth Century*, pp. 163–90.

When Aristotle says that things are said to be on various grounds, he is talking, here as elsewhere, of being in the sense of being real or existing. We, who speak in terms of uses of words, can say that there is the existential 'is', the copulative 'is' or 'is' of predication, and the 'is' of identity.

Aristotle does not draw this distinction, or say anything about the 'is' of predication or 'is' of identity, because he speaks in terms of things being called things, or said to be things. The only sense in which a thing can be said to *be* is an existential sense. When a thing is said to be red, the verb 'to be' is used predicatively, but the thing is not said to be predicatively: it is called red. Similarly if I say 'This chair is the chair I sat in yesterday', the 'is' is one of identity, but I do not say that the chair is in a peculiar, identitive, way: I call it one and the same as something, and Aristotle conceives it his task to say in what sense it and the chair I sat in yesterday are *one*, not in what sense it *is*. Whenever Aristotle talks about being, then, he talks about being in the sense of existing. Thus when he distinguishes being 'of itself' from being 'by virtue of concurrence' (*Met. Δ* 1017^a7 ff.), whilst he starts from predicative sentences like 'The man is just', his argument seems to be that because a man is just by virtue of concurrence, a just man exists by virtue of concurrence; and when he talks about being in possibility and being in actuality (*Met. Δ* 1017^b1–8), he is talking about that which *exists* in possibility, like half an undivided apple, and that which *exists* in actuality, like the undivided apple.

The basic error of the Eleatic monists was to think that things are said to exist in the first of the four ways described above, to think that whatever is real is real for the same reason, or because it accords with a single idea we have of existing (J. L. Austin suggested such an idea: like breathing, only quieter). What is Aristotle's own view? We have seen that in *Met. H.* 2 he represents material things as being said to exist on analogous grounds, but this account would hold only for things (like ice and thresholds) which are of roughly the same logical type. The Eleatic theory covered entities of every logical type, not only material things, but qualities, sizes, etc.; and Aristotle's opinion about these is that they are said to exist in the fourth of the ways distinguished above, in the way in which various logically heterogeneous entities are called healthy. (This, indeed, is what he normally has in mind when he says that 'things are said to be in many ways': *Met. Γ* 1003^a33, *Z* 1028^a10, *N* 1089^a7, etc.)

According to this view, there is a certain class of things, because of their relation to which, entities of other kinds are said to be real things, things which are. Aristotle's word for things of this logically primary class is *ousia*, which I translate 'reality'. It is usually translated 'substance', but it has none of the connotations of the English 'substance' or Latin 'substantia': it is simply the verbal noun of the verb *einai* 'to be', and the best translation would be 'being' if that were not also the translation of other parts of the verb. Aristotle normally uses the word precisely for whatever satisfies the criterion of logical primacy. Thus a dog is a reality, because it is the kind of thing which colours, sizes, shapes, etc. are the colours, sizes, shapes, etc. *of*. For the theory in general see *Met. Z* 1028^a10–31 and especially *Γ* 1003^b5–10: 'Many different things are all called real, but all with a relation to one source: some things are called real because they are realities, some because they are affections of realities, some because they lead to a reality, or are destructions, lacks, or qualities, or productive or generative, either of a reality or of other things which are relative to a reality, or because they are denials of a reality or of one of these other things.'

It will be noticed that Aristotle here says 'Some things are called real because they are realities'. We might say that some things are called healthy because they are certain bodily conditions: a temperature of 98·4 °F is healthy because it is part of a certain bodily condition, and (in a different way) a condition which enables us to enjoy food, sleep, and exercise is healthy because it is that condition. That which is called healthy because it *is* a certain bodily condition might also be called 'just what is healthy', 'just what health is', 'health itself'. In the same way, Aristotle seems to think that things like dogs and trees which are called real because they *are* realities are precisely what is real, are (cases of) reality itself (cf. 187^a8, *Met. Γ* 1003^b33). The notion of reality itself, of precisely what is real, plays a large role in the discussion of monism.

Since, according to Aristotle's theory, being real is not the same for realities, for quantities, and for qualities, we should ask the monist whether he means that there is only one sort of being real, e.g. only the sort of being real which belongs to realities, or only the sort which belongs to quantities, or only the sort which belongs to qualities. If so, there will be only realities, or only quantities, or only qualities (**185^a22–3**). And if the monist says that the only genuine kind of being real is that which attaches to realities like men and horses, does he mean that there is only one such reality? Or if he says that the only genuine sort of being real is that which attaches to qualities, like pale and hot, does he mean that there is only one such quality (**185^a23–6**)? Once the monist claim that there is only one thing is made precise, its absurdity is plain, especially if the one thing is something other than a reality (**185^a27–$^b5$**).

When Aristotle says (**$^a31$–2**) 'Nothing can exist separately except a reality; everything else is said of a reality as underlying thing', he does not mean that a reality like a man or dog can exist without qualities, or independently of any physical environment. He has in mind the logical point that colours are the colours *of* things like tigers and roses, whilst tigers and roses do not have to be in the same sense the tigers and roses *of* anything further.

Not only should we ask the monist what precisely it is which alone, he thinks, is real, but we should ask him in what sense he thinks it is one (**185^b5–7**), for there are three different grounds on which a thing may be called one thing. A lump of butter might be called one, if it is a continuous whole, i.e. not cut up into bits. At this point Aristotle notes that there are a number of problems concerned with wholes and parts (**185b11–16**). It is not quite clear what he has in mind when he talks (b14) of parts which are not continuous. Ross (following Pacius) suggests that the sheep which make up a flock would be non-continuous; the Greek commentators speak of organic parts in contrast with homeomerous bits; and it may be that Aristotle is thinking of 'parts of the form' of a thing (cf. *Met. Z* 1035b13–19) such as parts of the human soul. The last question is also obscure. Ross and Aquinas take it to be: 'Can organic parts like arms and legs each be identical with the whole, since if they are they will be identical with one another?' But why should anyone want to say that, e.g., the leg *is* identical with the man? Problems about parts and whole are considered in *Met. Z.* 10–11.

Second, a thing may be called one if it is indivisible. The word translated 'indivisible' can mean 'not in fact divided' (v. *De an.* III 430b6–7), and also 'indivisible' in various ways: spatially indivisible, logically unanalysable, indistinguishable in form. Here Aristotle seems to mean spatially indivisible, and hence unextended. If the universe is not extended, it will not have the qualities (185b17) which Parmenides wants to attribute to it, heat and cold (188a21–2) (if actual heat and cold must extend over an area or volume).

Third, we say '*X* and *Y* are one and the same' if the expressions '*X*' and '*Y*' are expressions for one and the same thing, like 'wine' and (in the sense in which it is sometimes used by poets) 'the grape' (**185b9**). Aristotle says (ibid.) that the account of the *ti ēn einai* of such things is the same. The phrase *ti ēn einai* is difficult. In the first place, it is unclear why Aristotle uses the imperfect form *ēn* instead of the present *esti*. My own view is that the construction is that of an unfulfilled conditional, with the *an* omitted, as is allowable, for euphony: 'what it would be' in place of 'what it is'. Second and more serious, it is unclear whether the *einai* is predicative or existential—whether the whole phrase conjoined with some further expression like 'a man' should be translated 'what it would be to be a man' or 'what it would be for a man to exist'. For a recent discussion see A. C. Lloyd, *Philosophical Quarterly* 1966, 258–67. Fortunately nothing turns on this question in *Phys.* I–II, but I have tried to preserve the ambiguity by translating 'what the being (of *X*) would be'.

If, says Aristotle, anyone maintains that all things are one in this third sense, i.e. that 'tree', 'dog', etc. are just different expressions for a single uniform reality, he will turn out to be saying the same thing as Heraclitus (**186b20**, see above, pp. 53–4) and making all knowledge illusory. The difficulties of the Heraclitean position, particularly in connection with judgements about what is good or bad, i.e. beneficial or harmful, are brought out by Plato in *Theaet.* 169–79, especially 172 a–b, 177 c–d, passages which Aristotle doubtless had in mind in writing 185b21–3.

This part of the argument is concluded with a historical note, **185b25–186a3**: not much is known of these 'thinkers of the more recent past', but we may meet ripples of their agitation in Plato, *Theaet.* 201 e ff. and *Soph.* 251 a–c. Aristotle's remark, **186a3,** that a thing may be one in possibility or in actuality, may be illustrated thus: a cake which has not been cut is one thing in actuality but several slices in possibility; bricks which have not yet been built into a house are several things in actuality but one thing in possibility.

In **186a4–22** Aristotle charges Melissus with two main absurdities. First, he thought it analytic that, if anything which comes into being has a beginning, anything which does not come into being has no beginning (186a11–13). Second, he thought that, if a process had a beginning at all, it must have a beginning in space; thus if a thing comes into existence, some bit must come into existence before the rest, and if it changes colour, the change must begin at some point and spread out from there (186a13–16). As the founder of formal logic, Aristotle is more outraged by the first mistake; but in fact it seems a harmless supposition that that which did not come into existence has no temporal beginning, and Melissus' paradoxical conclusions should rather be traced to his second mistake.

186a22–b12 is directed specifically against Parmenides, and contains the kernel of Aristotle's refutation of monism. The argument is to some extent anticipated by Plato, *Soph.* 244 b–245 d (where the awkward phrase *hoper hen* also appears: 244 c 1), and this may partly explain its compression.

Parmenides argues from the premiss, which, as we have seen, Aristotle considers false, that things are called real or existent for only one reason. This premiss, Aristotle proceeds to show, **186a25–32,** is not of itself sufficient to establish the conclusion that there is only one thing. Suppose that everything which is real is real for the single reason that it is pale (or white: v. textual note on 185a25), i.e. that to be real or to exist is to be pale. There can still be plenty of pale, and therefore plenty of real, things. We have no reason to suppose that they will form a continuous whole, and they will not be one 'in account', in the way in which clothing and raiment are one, i.e. one and the same thing. It is true that what it is to be pale is one thing; but there is a difference between what it is to be pale and that which is pale. Not that anything can exist except things which are pale. We are assuming that being pale is existing, and existence does not attach to anything which exists already, in the way in which knowledge of music might be grafted into an already existent man. But even though for a thing to exist is just for it to be pale, still pallor and that which is pale will be different.

Unlike Parmenides (a31–2), Plato in the *Sophist* does get as far as seeing this, but it is doubtful if he could have explained satisfactorily what the difference is. For Aristotle the difference 'in being' between pallor or 'the pale' and that which is pale, is that the former is a possibility and the latter a fulfilment of that possibility; and so long as a distinction of this sort is allowed, Aristotle's argument seems valid: even if there is only one possibility of existence, one kind of possible existence, there is no reason why there should not be a plurality of existing things.

Hence, says Aristotle, **186a32–4**, Parmenides must say, not merely that all things are called real for the same reason, but that if a thing is called real, that is because it is 'precisely what is' and 'precisely what is one'. These latter phrases, *hoper on*, *hoper hen*, have caused commentators perplexity. Some take Aristotle to mean: 'Parmenides must say that to call a thing real is to say that it is identical with the real and identical with the one.' This, however, makes Parmenides beg the question a little crudely, and the phrases are unnatural Greek for 'identical with the real' etc. *Hoper ti* in Aristotle normally means, I think (for a fair selection of examples v. Bonitz 533b39–534a23), 'precisely what is something' in the sense in which a certain bodily condition might be said to be precisely what is healthy.

That Parmenides should think that only precisely what is *f* in this sense can properly be called *f*, is likely enough. Plato in his middle period seems to have conceived the Form of *f* as precisely what is *f* in this way: the Form of large is that which in the primary sense is large, it is the large itself, that by virtue of their relation to which other things are called large. Cf. also Aristotle's criticism of the Pythagoreans, *Met.* *A* 987a21–5.

The trouble with interpreting *hoper on* like this is that, if we do, the premiss will still not be strong enough to yield a monist conclusion. We might say that the only thing which can properly be called healthy is a certain bodily condition: it will not follow that there is only one healthy thing. What Parmenides must do is combat any distinction such as we might try to draw between what it would be for a man to be healthy, and a particular healthy man's bodily condition, between health as a possible physical state, and your health and my health. If to know what the word '*f*' means, is to know what it would be for a thing to be *f*, we might put the matter thus: Parmenides must say that the only thing to which any word, including the word 'real' or 'existent' applies, is its meaning. If that is so, then since, according to him, 'real' or 'existent' has only one meaning, there is only one existent thing.

Plato is perhaps trying to propose this premiss for Parmenides in the obscure *Soph.* 244 d 11–12; is Aristotle trying to do the same here? The argument which follows is easier, I think, to understand, if we make him give Parmenides the weaker premiss. As Mlle S. Mansion observes

('Aristote, critique des Éléates', *Revue philosophique de Louvain*, May 1953, 177) the lines **186a34–b4** become more intelligible if we transpose (at least in thought) the sentence 'Precisely what is, then, will not be something which belongs to something else' to the beginning. If, says Aristotle, precisely what is real *were* something which belongs to something else (as precisely what is healthy, the physical state, belongs to Socrates), that to which it belongs would not be real; for it would be different from that which is properly called real (as Socrates is different from his physical state). Parmenides cannot say that it (sc. the thing to which 'precisely what is' belongs) is a *kind* of real thing (as Socrates is a *kind* of healthy thing, *hygieinon ti*), unless he allows that things can be called real for different reasons, which he does not.

Having argued that being real cannot supervene on or belong to anything other than precisely what is real, Aristotle proceeds to claim (**186b4–11**) that precisely what is real might just as well be called unreal as real, since all positive determinations must be denied of it. His point is comparable with the one Hegel makes at the beginning of his *Logic*, that mere being could just as well be called nothing. Suppose that what is real is also pale; and suppose, what is necessary if our notion of that which is real is not to remain vacuous, that to be pale is not just what is real (as to be 98·4 °F, perhaps, is just what is healthy): then just what is real will be unreal, for it is pale, and pale means not real. This argument may at first seem invalid: Aristotle seems to be saying: 'Because to say that a thing is pale is not to say that it is real, it is to say that it is not real'—which is plainly fallacious. However, according to the theory of meaning which Parmenides has invoked, to say that a thing is *f*, is to say that it is just what is *f*. Hence to say that a thing is pale, is to say that it is just what is pale. Now we assumed that just what is pale is different from just what is real; and it was shown in a34–b4 that being real does not belong to anything except just what is real. From this it does follow that if a thing is pale, it is not real.

Finally, **186b11–12**, Aristotle suggests that the Eleatics might try to escape this conclusion (and hence the conclusion that all positive determinations must be denied of that which is real) by saying that 'pale' also means just what is real. The idea is perhaps that just what is real is also just what is pale, rather as the same rod might be the standard kilogramme and also the standard metre (just what does weigh a kilogramme, just what is a metre long). If this is what Aristotle has in mind, his reply is probably that in that case there must be both reality and quality (as in the case of the rod, there would have to be both weight and length), which is what the Eleatics originally (185a23) denied.

This concludes the central argument against Parmenides. How effective is it? For it is here—and not, as L. Taran, *Parmenides*, p. 284, declares, in *Met. A* (see 986b30–1)—that we must look for Aristotle's formal critique of Parmenides. Although Parmenides would not have seen the difference between the stronger and the weaker forms of his premiss, I think it likely he held more to the weaker. He needs nothing more than the weaker to say that anything different from being itself is non-existent, which seems to have been the *nervus probandi* of his monism. If he did argue from the weaker premiss, Aristotle should have pointed out that it is too weak; instead, Aristotle seems to consider it strong enough. However, Aristotle's criticism that Parmenides is unable to assert anything positive about his one reality is persuasive, and is reinforced by the argument, 186b14–35, that the one reality is also logically unanalysable.

Whether Aristotle's argument would succeed against a monist of greater sophistication, like Spinoza, is doubtful. Spinoza allows, what Parmenides does not, that things can be called real for more than one reason. Modes, like dogs and the human mind, exist because they supervene on, or modify, the one reality; the one reality exists because it is just what is real. Further, Spinoza seems to have adopted the stronger monist premiss. His one reality is not an instance of what it is to be real (as Socrates' bodily condition is an instance of what it is to be healthy); rather it is what it would be for anything to be real itself. (The point is not made in so many words, but seems to underlie the crucial *Ethics* I. viii. sch. 2, that, whilst there is a difference between what a term like 'triangle' means and that to which it applies, a substance is by definition something over which this distinction cannot be drawn.) To protect us against this line of argument Aristotle would have to emphasize that the distinction between that which is because it is a reality and that which is because it is related to a reality, is quite separate from the distinction between that which is in possibility and that which is in actuality.

There remain two other arguments, also of limited efficacy. The first, **186b12–14**, is that a monist cannot allow what exists to have magnitude, since whatever has magnitude can be divided, and what exists will therefore possess parts, each with its own existence. The second argument, **b14–35** is variously interpreted. Some (e.g. Ross, Gershenston, and Greenberg) suppose that Aristotle is making the Eleatics grant that there really are entities like men in the universe, and asking whether they are analysable. Ross takes the final obscure question 'Does the universe, then, consist of indivisibles?' (b35) to mean 'Must we then suppose that such entities are unanalysable?' I prefer the view of Philoponus that man is being used as an illustration of a formal point, like pale in a26–31, and that Aristotle is contending that, if what it is for Parmenides' one reality to be real is analysable in the way in which what it is to be a man is analysable, there will be a plurality of things, each of which is what it is to be real; and I understand the final question as: 'Is the universe, then, since it cannot be analysable, an aggregate of unanalysables? Has monism turned into a form of atomism?' Whichever way we take the argument, it might not much have disturbed Parmenides. He would have been most irresponsible to allow the existence of things like men, and he is under no obligation to make existence analysable in the way in which being a man is analysable. He could have said, and perhaps does in DK 28 B 8, that the being of the one reality is itself simple and one, but appears differently according as it is regarded from different points of view.

Aristotle concludes, **187ᵃ1–11**, with a note to the effect that certain thinkers gave in unnecessarily to two celebrated Eleatic arguments, the argument from the meaning of the word 'is', which we have been considering, and Zeno's arguments that the supposition that reality is divisible leads to intolerable paradoxes. (For a discussion of the divisibility of matter see *De gen. et cor.* I. 2, and for a reply to Zeno's paradoxes about motion, *Phys.* VIII. 8). Ross and others think that Aristotle has in mind Zeno's argument that the many would have to be both infinitely great and infinitely small, both limited and unlimited in number (cf. Plato, *Parm.* 127 e), but I think it more likely that by the dichotomy Aristotle means the paradox of the stadium: v. *Phys.* VI 239ᵇ18–22. There may also be discussion about who these thinkers are. *Met. N.* 2, especially 1089ᵃ2–6, suggests that they are Platonists, but Ross points out that Plato was aware that 'that which is not' can be understood as that which is not something definite, i.e. that not being can be analysed as difference (*Soph.* 258–9), and that *De gen. et cor.* I 324ᵇ35–325ᵃ32 tells strongly for the view that Aristotle is here referring to the atomists. I agree; and if we were to read ἔνιοι γὰρ for ἔνιοι δ' in 187ᵃ1, the question 'Does the universe then consist of indivisibles?' could be taken with what follows, and would give no more difficulty.

CHAPTER 4

In this chapter Aristotle reviews theories which had been held by the Presocratic physicists about the principles of things. He divides the physicists into two groups (**187ᵃ12–26**). Some held that there is intrinsically uniform matter, e.g. water, air, or fire—'the three' of ᵃ13—and that this constitutes different things according as it is densified or rarefied; others held that matter is intrinsically diverse or multiform. Aristotle says that Plato belongs to the first group—with what justice will be discussed below, pp. 84–7; more obvious members of it would be Thales, Anaximenes, and Heraclitus, cf. *Met. A* 984ᵃ2–8. To the second group belonged Anaximander and, more formidable, Empedocles and Anaxagoras. Empedocles held that there are four 'roots' from which everything arises (DK 32 B 6), fire, air, earth, and water, which came to be known as the four elements (v. textual note on 187a26). Anaxagoras is said to have posited an unlimited number 'both of homeomerous things and of opposites' (a**25–6**). A thing is called homeomerous if the same description which applies to it applies to parts of it. Thus bone, or a piece of bone, is homeomerous, because bits of bone are bone, and half a piece of bone is still a piece of bone. Aristotle may be saying that Anaxagoras held that there were infinitely many kinds of homeomerous stuff as well as infinitely many pairs of opposites, hot–cold etc.; or he may (perhaps more forcefully) be saying that besides infinitely many pairs of opposites, Anaxagoras posited infinitely many particles or 'seeds' of each kind of homeomerous stuff (cf. DK 59 B 4).

The significance of this grouping appears from *De gen. et cor.* I 314b1–6: 'Those who derive everything from a single type of matter must make coming to be and ceasing to be alterations. The underlying stuff remains one and the same, and what is like that is said to be altered. For those who posit several kinds of matter, alteration will be different from coming to be. Coming to be and ceasing to be will occur when things come together and separate.' That is, as the rest of the chapter shows, those who make the matter of things in itself uniform can and must allow qualitative change. When an egg becomes a chicken, or water in a kettle becomes hot, the underlying stuff alters. Those, in contrast, who make the matter of things in itself diverse, in itself determined by qualities like hot, cold, pale, dark, wet, dry, soft, hard (314b18–19), cannot consistently allow that there is alteration, since alteration is precisely change in respect of such qualities.

By alteration we normally understand a change such as a tomato undergoes when it turns from green to red: in such a case there is a definite, identifiable thing, a tomato, which remains throughout the change. A man like Empedocles cannot allow alteration in this sense, at least in respect of basic qualities like hot and cold, because these determine his elements; there is no concept under which he can identify a thing which changes from hot to cold throughout the change. From this alone it does not follow that Empedocles must do away with qualitative change altogether, and say that what appears as a case of rise in temperature is really a case of hot stuff coming along and/or cold stuff departing. There is still the possibility that the cold stuff changes into hot stuff in such a way that the change is a ceasing to exist of the cold stuff and a coming into existence of the hot. However, this possibility Empedocles will not allow (314b23–5), on the ground, presumably, common to all the other physicists, that that which comes into existence must do so either out of nothing or out of what exists already, and neither is possible (191a28–31, cf. 187a32–5). Hence although Anaxagoras (187a30) and Empedocles (cf. 189a24–6) did in fact allow qualitative change, Aristotle is right to accuse them of being inconsistent (315a3–4).

The view of the first group of physicists may at first seem more attractive: there is a quantity of matter in the universe which neither comes to be nor passes away, but merely changes in quality—in colour, temperature, etc., or, perhaps better, in shape, state of motion, and the like. It is generally accepted that Aristotle's own view is of this kind—that he posits a single, universal, indeterminate substratum for all change. Such a line, however, is not free from difficulty. It involves what Aristotle calls a 'separation' of matter and qualities, the matter becomes embarrassingly unknowable, and the qualities slide in and out of the actual world in a way which raises just those questions about coming to be and ceasing to be that the theory of a permanent substratum was designed to evade. The traditional view that Aristotle is none the less committed to this line will be challenged below.

In the present chapter Aristotle is mainly concerned with the second group of physicists. His arguments against them are on the whole straightforward. The clause **187b30**, which I translate 'there will always be some quantity smaller than any yet yielded', is literally: 'it [sc. the yield at any time] will still not exceed some magnitude in smallness'. I take Aristotle to mean: for all x, x is a yield implies there is a y such that x is not smaller than y. Others take him to mean: there is a y such that for all x, x is a yield implies x is not smaller than y. The latter would be a better premiss for Aristotle's argument, but it is hard to see how he could establish it. When in **188a14–15** Aristotle says that there is a sense in which clay does not divide into clay, he is probably thinking of it as dividing into earth and water (cf. Plato, *Theaet.* 147 c). On bricks and walls (**a15–16**) cf. *De gen. et cor.* II 334a19–b2.

CHAPTER 5

In this chapter Aristotle offers two arguments for the view that the principles of physical things are opposites. One (**188a19–30, 188b26–189a10**) is an argument from authority or *ex consensu sapientium.* With the remarks on Democritus (**188a22–6**) compare *Met. A* 985b13–19. The atoms were made of homogeneous stuff, but they constitute different things according as they differ in shape, posture, and order (cf. Locke, *Essay* II. viii. 10–14). The fanciers of odd and even and love and strife (**188b34**) were, respectively the Pythagoreans (v. *Met. A* 986b15–19), and Empedocles. For the distinction between things known by perception and things known by means of an account (**189a4–8**) cf. Plato, *Politicus*, 285 d–286 a.

The other argument, **188a30–b26**, is based on consideration of the *logos* (a31). Ross takes this as meaning 'from a consideration of the argument' and cites as a parallel *De gen. et cor.* I 325a14, where Aristotle speaks of philosophers who say one ought to follow the argument. This passage is not in fact a good parallel, because Aristotle is apparently quoting a well-known slogan, and the argument in question is a famous one. Better for Ross's interpretation are Plato, *Rep.* I 349 a 4–5, *Laws* V 733 a 6–7, etc. However, another passage in Plato, *Phaedo* 99 e 4–100 a 2, suggests that 'considering *logoi*' is simply considering speech, or things said, and this passage is the more deserving of attention here, because the argument which follows is foreshadowed by *Phaedo* 70 c–72 d. I have tried to leave the matter open by translating 'from logical considerations'; the phrase seem to me akin in meaning to 'logically' in *Met Z* 1029b13 (cf. *Met. A* 987b31–2 with *Λ* 1069a28), and the logical considerations adduced there turn out to concern the way we speak.

Aristotle says that it is not a matter of chance what comes to be out of what, but a thing always comes from its opposite or something in between. This is not an empirical doctrine to the effect that the universe is regular; it is the purely logical doctrine that change is within definite ranges. We would say that a thing changes from being red to being blue, or from being round to being elliptical; we would not say that a thing changes from being red to being elliptical, or from being round to being blue—though of course something round which changed to being elliptical might also have happened to be red, cf. **188a34–6**. This seems to be a sound point, and one way of understanding an Aristotelian 'kind of thing' or category (189a14, b24–6) is as a range within which things may change.

In taking this line, Aristotle diverges both from the Presocratics and from Plato. He differs from the Presocratics, in that whilst they made everything come to be out of the same opposed principles, either dense and rare or cold and hot or the like, he makes things come to be out of different but analogous opposed principles. In so doing, he removes the discussion from the sphere of empirical to the sphere of philosophic inquiry. And his insistence that pale does not come from just anything other than pale but from the opposed state, is probably directed against Plato, who in the *Sophist* construes 'that which is not *f*' as 'that which is not identical with *f*', so that it covers not only whatever is opposed to *f*, but also things which have nothing to do with *f* at all: see 256–9, especially 259 b. If Aristotle were asked whether Plato is not as competent as himself to remove the difficulty about coming to be experienced by the Presocratics (see above, p. 64) this is probably one of the points he would make.

The outline of 188a30–b26 is fairly clear, but there are a couple of points left in some obscurity. First, the nature of the opposition. Pale and dark, hot and cold, are indefinite opposites: neither 'pale' nor 'dark' is the expression for a definite colour—things are called pale and dark relative to some norm and, e.g., a pale Sicilian may be darker than a dark Swede. On the other hand, the arrangement of bricks in a house is something definite, and whilst the state of 'being arranged not thus but otherwise' may be called indefinite (cf. *De int.* 16a30–2), the two are opposed, not like pale and dark or hot and cold, but rather like correct and incorrect or hitting and missing. It will appear in chapter 7, but hardly appears here, that the opposites which are principles are opposed in this latter way.

Second, it is unclear whether the opposites are entities the correct expressions for which would be abstract, like 'pallor', 'knowledge of music', or concrete, like 'pale thing', 'thing which knows music'. Aristotle uses the neuter adjective with the definite article, which may be taken either way. We shall have to settle this point too when we come to chapter 7.

For the idea (**188b23–5**) that particular colours are 'out of' i.e. compounds of pale and dark cf. *De sensu* 3, Plato, *Tim.* 67 d–68 d.

CHAPTER 6

In this chapter Aristotle argues that whilst it cannot plausibly be held that the principles of physical things are less than two or more than three in number, there are reasons for thinking they may be as many as three. A hasty reading might make us think that Aristotle is arguing that, besides the opposites of the sort identified in chapter 5, we must always suppose that there is a third factor underlying them. In fact, he is careful not to be so dogmatic. Whereas he usually describes even the most questionable points he makes as clear or plain (*dēlon, phaneron*), here he uses carefully guarded language: there is an argument for positing an underlying thing, 189a21–3, b17–18; people might feel difficulties otherwise, a22, 28; if anyone accepts certain arguments, he must say so and so, a35–b1; but in the end, whether we are to posit underlying things remains a very difficult question, b29. The truth is that Aristotle is presenting a mild antinomy: the arguments that the principles are opposites suggest that there are two in number, but there are also arguments suggesting they must be as many as three. Chapter 7 is intended, among other things, to resolve the antinomy (v. Aristotle's summary of the whole discussion, 191a15–19).

Aristotle begins by rehearsing the arguments against allowing either one principle only, or an unlimited number (**189a11–20**). Among the difficulties about positing an unlimited number he includes the fact that there is 'only one opposition in each kind of thing, and reality is one such kind (**189a13–14**). 'Kinds of thing' were explained above

(p. 66) as ranges within which things may change. (In fact, a category like quality is a set of such ranges, but the point that there is only one opposition to one kind holds only for individual ranges: cf. *De An.* II 422b23–31). We also saw that some ranges at least could be characterized by pairs of indefinite opposites like tall and short, pale and dark. This account of a 'kind' does not apply well to realities. In the first place, we may think that ranges within which changes occur are ranges within which realities change, so that realities cannot themselves constitute a range. However, if we take a 'kind' as a range in which things may differ, reality might be one such range. Two things can differ in that one is a dog and the other a tree, and we might call this a difference in reality. Second, Aristotle elsewhere, e.g. 189a32–3 and especially *Cat.* 3b24–7, says that realities do not have opposites. Perhaps, however, we may understand him here as adopting for the moment the position of a philosopher who thinks that there is one kind of underlying stuff, which constitutes different things according as it is modified in different ways, according as it is denser or rarer, or hotter or colder, or the like. In that case, his argument against a plurality of opposites may be something like this. Suppose a portion of stuff constitutes a tree because it is densified a certain amount. If we then say that another portion constitutes a horse because it is heated a certain amount, we shall not be able to say that these two portions differ in that one is a horse and the other a tree, for the range of temperatures is different from the range of densities. If a tree is a reality, and to be a tree is to be determined by one pair of opposites, say dense and rare, to be any other sort of reality must be to be determined by the same set of opposites. The argument, though Aristotle recurs to it **189b22–7**, is not very convincing: we might say of realities in general what Aristotle himself says of hands and feet (*Met. H* 1042b28–31), that they are determined by a number of different sorts of differentiating feature taken together.

If the principles are neither one nor unlimited in number, can they be only two? Aristotle offers three grounds for thinking not. First, if the opposed principles are taken as properties, like density and rarity, or love and strife (poetic names for combination and dissolution) they cannot act on one another, but there must be some third thing on which they act (**188b22–6**). Aristotle is unquestionably treating the opposites as properties, not things, here, but he is saying how people might be led to posit an underlying thing, not speaking for himself.

Second, no opposite seems to be the reality of anything, and a principle should not be something said *of* something else: that of which it is said will be prior to it and more of a principle (**189a29–32**). Aristotle might here be taking the opposites as properties, and saying that we call a temperature or a density real if it is the temperature or density *of* something; but I think that what he has in mind is rather that 'dense' and 'hot' are not expressions for particular things in the way in which 'a dog' or 'a tree' is an expression for a particular thing, and that we do not call dense or warm a cat's fur, but rather call a cat's fur warm and dense. Whichever way we interpret the passage, it should not be taken as a dogmatic assertion that reality or substance is matter. There is no reason to think that by a reality Aristotle means anything other than the sort of thing he elsewhere calls a reality: a plant, animal, or artefact.

Third, we say—and 'we' here means the Lyceum, not the man in the street—that a reality has no opposite. Now hot and cold, dense and rare, and the like, are opposites. If, then, they constitute realities, things which are not realities will be prior to things which are. But a reality is defined as that which primarily is, and by virtue of their relation to which other things are said to be. Hence opposites like dense and rare cannot be the only things which are principles in the sense of constituents of physical things (**189^a32–4**). Here also Aristotle means by a reality something like a horse, tree, or statue, and is not conceiving the opposites specifically as properties. These last two arguments do not establish that there is an underlying thing over and above the opposites, but they do show, by emphasizing that physical things, the things whose principles we are seeking, are realities, that no indefinite opposites, like hot and cold or dense and rare, can qualify as principles.

In **189^a34–$^b16$**, Aristotle shows how these considerations had influenced or might influence the natural scientist. The passage is perhaps slightly ironical. The recent thinkers of **$^b15$** are Plato and his followers—their opposites were great and small—and $^b11$–16 constitutes a dig at them. Aristotle's own treatment of the matter in chapter 7 will shew that there are in fact no scientific conclusions to be drawn at all.

Finally, in **189^b16–28**, Aristotle argues that the principles are not more than three in number. If there are four principles, i.e. two pairs of opposites, then either each pair will need a further principle as underlying nature, in which case we shall have not four principles but six; or (as I understand the next clause, **$^b22$–3**) if the principles in each pair can produce things out of one another without a third principle, one of the pairs will be redundant, presumably because the pairs are 'by analogy the same' (cf. 188^b37–189^a1): if, e.g., horses are produced by hot and cold, and trees by wet and dry, the principles of each, though different, will be analogous. On this interpretation, the next point, **$^b22$–7**, follows naturally: if horses are produced by hot and cold, and trees by wet and dry, and the one pair cannot be reduced to the other (as perhaps pale and dark, or colours, can be reduced to rough and smooth, or textures—cf. *Met. Z* 1029^b21–2), horses and trees will not be in the same range or 'kind'. Others, e.g. Ross, take $^b22$–3 to mean 'if the opposites in each pair will serve as underlying things for the other, but this is perhaps too complicated a thought for the words to carry.

CHAPTER 7

This chapter, in which Aristotle puts forward his own (189b30) account of the principles of physical things, is generally agreed to constitute his formal introduction of the notions, fundamental in his thinking, of matter and form, and he himself seems to refer us to it for detailed treatment of these notions in *De gen. et cor.* I 317b13, II 329a27, *Met. M* 1076a8–9, etc. Unfortunately it contains ambiguities. Aristotle uses neuter adjectives with the definite article, which as we have seen (p. 67) can be understood in two ways, and he makes much play with the verb *gignesthai* (e.g. 190a28–31), which can mean either 'to become' or 'to come into existence'. I have tried to preserve these ambiguities in my translation; to appreciate them fully the reader should remember that the phrase I translate 'the ignorant of music' could be used for the state 'ignorance of music'.

The general view of commentators is that an Aristotelian form is an entity the natural expression for which is an abstract noun or equivalent phrase, like 'knowledge of music', 'sphericality', 'what it would be to be a man'. (It is because of this that they find obscure the argument of *Met. Z*, that forms are the entities with the best claim to be called realities in the sense (v. 1028b36–1029a9) explained above, p. 56; for Aristotle constantly says that only a particular thing, 'a this thing here', can be a reality, and it is hard to see how something like man-ness could be a this thing here, or a thing which colours, sizes, etc., are *of*.) If this is right, then since the three factors involved in any case of change are the matter, the form, and the lack (190b23–9, 191a12–14, etc.), the factors involved in Aristotle's case of the man who learns music ought to be the man, ignorance of music, and knowledge of music. In support of the view that these are the factors he is really trying to elicit, the following passages may be cited. In 190b15 we have abstract nouns for the *terminus a quo* factor, 'shapelessness' and 'formlessness', and in b28 for the *terminus ad quem*, 'arrangement', 'knowledge of music'. Similarly in chapter 5 Aristotle speaks of opposed dispositions (188b11) and uses the abstract nouns 'disunion' (b14), 'shapelessness' (b20), and the verbal phrase 'the not being put together but dispersed thus' (b18–19). And in *Met. Λ* 1070b28–9, in a context similar to the present, he gives the examples health, disease, and body, and form, such and such a disorder, and bricks. Further, Aristotle's generic expression for the opposite from which change takes place is an abstract noun, 'the lack', *sterēsis*.

Nevertheless, translators and commentators seem agreed that the factors distinguished when a man learns music are not the man, ignorance of music, and knowledge of music, but the man, the thing which is ignorant of music, and a thing which knows music. If Bekker's reading in 189b35 *to mē musikon ti* is right (in my translation, with reluctance, I follow Ross's text with the *ti* omitted), Aristotle says this unambiguously. Even if it is wrong, this is still the most natural way of understanding his words, and is confirmed by *mousikos* instead of *mousikon* in 190^{a}7. If this is so, however, it becomes questionable whether an Aristotelian form is, after all, an entity the natural expression for which is an abstract noun. And if a concrete expression is just as natural or more so, doubt is cast on the whole traditional interpretation of Aristotle's teaching on matter and form. The relation of matter to form is traditionally construed, I think, as a kind of thing–property relationship, like that of a man to knowledge of music, or of bronze (see below) to sphericality; if the authentic model for the matter–form relationship is that of man to thing which knows music, or of bronze to a sphere, the relationship must be construed differently. Evidence telling for abstract expressions has been given above; evidence telling for concrete expressions is, I think, much stronger, and since the issue is important, I give it in some detail. It suggests that the matter–form relationship is that of constituent to thing constituted (cf. D. Wiggins, *Identity and Spatio-Temporal Continuity*, p. 48).

1. In chapter 5 Aristotle reckons among *termini ad quos* of change, houses and statues (188^{b}17). It is true that he also says that such things are all arrangements or compositions (^{b}20–1), but that only allows us to gloss abstract expressions elsewhere with concrete ones: Aristotle may be thinking of arrangements and compositions, not as things added to, but as things constituted by, bricks, bronze, and the like; indeed, he says in *Met. H* 1043^{b}5–6: 'The syllable does not consist of letters *and* composition; the house is not bricks *and* composition.'

2. In the formal explanation of the 'underlying nature' in 191^{a}8–12, Aristotle says: 'As bronze stands to a statue, or wood to a bed, or the formless to anything else which has a form, so this stands to a reality.' If this passage were taken by itself, it might be held that Aristotle thinks 'a bed', 'a statue' are natural expressions, not for forms, but for things which have forms; but in conjunction with (3) and (4) below it suggests that 'a bed', 'a statue' are themselves acceptable expressions for forms.

3. In the formal classification of causes, 195^{a}16–21, Aristotle says 'Letters are the cause of syllables, the matter of artefacts, fire and the like of bodies, the parts of the whole, and the hypotheses of the conclusion, as that out of which; and the one lot, the parts and so on, are causes as the underlying thing, whilst the other lot, the whole, the composition and form, are causes as what the being would be'. It is hard to understand Aristotle otherwise than as implying that syllables, artefacts, bodies, and wholes *are* forms, and that the matter–form relation is that of constituent to thing constituted.

4. In the formal explanation of the notions of possibility and actuality, *Met. Θ* 1048^{a}36–^{b}6, Aristotle says: 'We need not seek a definition for every term, but must grasp the analogy: that as that which is actually

building is to that which is capable of building, so is that which is awake to that which is asleep; and that which is seeing to that which has the eyes shut, but has the power of sight; and that which is differentiated out of matter to the matter; and the finished article to the raw material. Let actuality be defined by one member of this antithesis, and the potential by the other.' (So Dr. M. Hesse, in 'Aristotle's logic of analogy', *Philosophical Quarterly* 1965, p. 335.) If we take this passage with the last two cited, and with Aristotle's statement in *Met. H* 1045^b17–19 'The last matter and the form are one and the same thing, the one in possibility and the other in actuality', we must surely understand that wood and a bed, bronze and a sphere, and the like, are examples of matter and form. And if we do not understand bronze and a sphere to be examples of matter and form, we shall find it extremely hard to understand 1045^b17–19, or the rest of the chapter, *Met. H* 6, in which it occurs.

5. Other less formal passages in the *Metaphysics* point to the same conclusion. In *Met. Λ* 1069^b36–1070^a2 Aristotle says: 'Wherever there is change, *something* changes, by the agency of *something*, to *something* . . . The thing *which* changes is the matter; the thing *to which*, the form.' In 1032^a13–19 he says: 'Whatever comes to be, comes to be through the agency of something, and comes to be out of something, and comes to be something . . .; that out of which it comes to be, we call the matter . . .; that which it comes to be, is a man or plant or the like.' See also 1033^a10–12, 24–8.

6. The difficulty mentioned above, about the line of thought in *Met. Z–H* which equates reality or substance with form, disappears if 'a man', 'a sphere' are expressions for forms. In this connection we should notice that, though Aristotle has the abstract noun 'sphericality' or 'roundness' (*stroggulotēs* 1035^a14), whenever in *Met. Z* or *H* he speaks of the form of a sphere or circle he calls it 'the sphere', 'the circle'. Similarly the form of a house is called either simply 'a house' (1033^b20) or 'a shelter' (1043^a33, cf. *De an.* I 403^b4). On the other hand, when he wants to talk about things consisting of matter and form, he uses expressions like 'bronze sphere', 'clay statue' (1033^b9–16, 1035^a26–34, 1045^a26–9).

7. In *De gen. et corr.* I 321^b19–34 Aristotle claims that when a living thing grows, it is strictly the form, not the matter, which gets larger. A form here is clearly a thing constituted, e.g. a cucumber, not cucumber-ness.

8. A similar question arises over Plato's forms, and many now hold it best to use concrete expressions for them. Aristotle himself regularly does so, even though he claims that Plato 'separates' forms from matter in an improper way (e.g. 193^b35 ff.) It would be surprising, then, if the proper expression for an Aristotelian form were abstract. Rather, when Aristotle does use an expression like 'what it would be to be a horse' for a form, we should suppose he does so because he thinks that what it

would be to be a horse is the same as a horse in the sense of that which this flesh and bone constitutes (cf. *Met. Z* 1031^{b}29–31); where we have a bronze sphere, he would say, the bronze does not so much possess as constitute an instance of sphericality (cf. *Met. H* 1045^{b}7–22).

This evidence, I think, establishes a strong presumption that Aristotle's matter–form distinction is primarily (for qualifications see below, pp. 95–6) a distinction between constituent and thing constituted, between what a thing is made of and what that of which it is made makes or constitutes. Now how would such a distinction naturally be presented in Greek? The Greek verb for making, *poiein*, cannot suitably be used in connection with things which are not manufactured, like plants and animals; and would not be natural even to translate 'make' in 'These bricks make' or 'make up' 'a house'. Aristotle has a passive verb *sunistasthai* for 'to be constituted', but the active voice of this verb would not be used like our active 'constitute'. The natural Greek verbs in this context are *einai* and *gignesthai*, 'to be' and 'to come to be'; and the distinction would naturally be expressed as the distinction between that out of which a thing comes to be or is, and that which comes to be or is out of this. This is in fact the terminology Aristotle uses in this chapter, and he encounters precisely the difficulties to which it gives rise. There are more things than one 'out of which' a thing may be said to come to be, and it is in fact only artefacts like a statue that are made of or constituted by that out of which they would naturally be said to come to be.

Aristotle's account is put forward as a solution of the antinomy reached, as we saw, at the end of chapter 6, about whether the principles are two or three in number. The outlines of the solution are fairly clear. It is indeed true that in all cases we must suppose an underlying thing, but the underlying thing is not a third factor over and above the opposites: it is the same thing as one of the opposites, viz. that from which the change takes place, but under a different description. Aristotle tries to establish this separately for two classes of things, things the change to which is an alteration, and things the change to which is a coming into existence.

The argument begins with the analysis of a case of alteration: a man learns music. Aristotle here distinguishes three 'simple' factors, which seem, as we have seen, to be the man, the thing which is ignorant of music, and a thing which knows music (**189^{b}34–190^{a}3**). The man and the thing which is ignorant of music are the same thing under different descriptions. Under the former description this thing remains, and under the latter it does not; and it is under the latter that a thing which knows music comes to be out of it (**^{a}5–23**). The last point reflects a way of speaking corresponding to our 'From (being) a thing which is ignorant of music, the man comes to be a thing which knows music' or 'From green, the tomato comes to be red'. It is not in this sense of 'out of' that the material factor is 'that out of which', and in fact the material factor in such cases of alteration has to be characterized as that which the *terminus ad quem* is *not* out of, and which remains throughout the change. This first part of the argument is fairly straightforward, but some details need attention.

In what sense is a thing which knows or is ignorant of music simple, when a man who knows or is ignorant of music (**190a4–5**) is not? An answer might be extracted from the difficult *Met. Z* 4, especially 1029b22–1030a5, 1030a29–32. A man can constitute a thing which knows music, and though constituting a thing which knows music is different in important ways from constituting a man or tree, it is still one thing, and we can say what it would be to constitute or be a thing which knows music. It is not the same with a man who knows music. It is incoherent to talk of a man constituting a man who knows music, and for anything other than a man (e.g., perhaps, a quantity of flesh and bone) to constitute a man who knows music, is not one thing but two, for it is one thing to be a man, and another thing to know music. Hence a thing which knows music is simple in a way in which a man who knows music is not. The point may be more acceptable, if a thing which knows music is understood not as a possessor of, but rather as an exemplification or instance of, knowledge of music. An instance of knowledge of music is clearly more 'simple' than a man who knows music, and if an instance of knowledge of music is simply a thing which knows music as such, then a thing which knows music may also be called simple.

Aristotle says that the thing which is ignorant of music and the man are one 'in number' but two in form or account (**190a15–16**). By 'in number' he means 'in reality' or 'in fact' (cf. *Phys.* VIII 262a21, 263b13, *De an.* III 427a2, etc.), but why does he use the phrase? Perhaps because where we can say 'in reality, as distinct from in form or nature', we can also say 'in number'. Thus if I own three sheep, they are three in number and reality, but one in nature, form, account. (In general, it is fulfilments or realizations of possibilities which are numbered: we might talk of three performances of the same play, three makings of the same journey.)

In what sense does the underlying thing remain in alterations (**190a10–11**, and see textual note)? A man who becomes a musician does not thereby cease to be a man, but there seems to be more to it than that. In a**24–6** Aristotle says that a thing is usually said to come to be out of the factor which does not remain, but sometimes out of the factor which does; for instance we say that a statue arises out of bronze, not that bronze becomes a statue. This is explained in *Met. Z* 1033a13–18 (cf. also *Θ* 1049a19–20) and most interestingly, perhaps, though the authority of the book is questionable, in *Phys.* VII: 'When something is shaped or moulded to completion, we do not say that it is that out of which it comes: thus we do not call the statue bronze or the candle [so Ross ad loc.] wax, or the bed wood, but, by a modification of those expressions, we call them brazen, waxen, wooden. But of that which has been affected and altered, we do speak so. We call the bronze and the wax liquid and hot and hard, and not only that, but we also call the liquid and the hot bronze, speaking of the matter in the same way (?) as the affection. So if, when the change is in respect of shape and form, we do not call the thing which comes into being that in which the shape is, whilst when the change is in respect of affections and an alteration, we do, it is clear that comings into being are not alterations' (245^b9–246^a4). That is, whilst, when a man becomes a musician, the thing which knows music can correctly be called a man, a statue cannot correctly be called bronze, but only brazen. (Or, perhaps: we can refer to a musician as 'that man', but we cannot refer to a statue as 'that bronze', but only as 'that brazen thing'.) And this asymmetry seems to occur because the case of a man's becoming a musician is one of alteration, whilst a statue's arising out of bronze is the coming into existence of a reality. At any rate, Aristotle here classes artefacts like statues and houses as realities, for they come into existence (190^b5), and only realities come into existence ($^a32$–3). This brings us to the part of the argument dealing with realities, **$^a31$–$^b9$.**

With realities too, says Aristotle, there is always something underlying, from which the reality arises, as plants and animals arise from seeds; and he then treats artefacts as being on all fours with plants and animals. There are two difficulties we may notice about his account.

First, in *De gen. et cor.* I 4, Aristotle raises the question whether there is such a thing as coming into existence over and above alteration and, if so, how the two can be distinguished. He answers as follows: 'If some affection [*pathos*, an extremely general word] in that which has passed out of existence remains in that which comes into existence, as transparent and cold do when air turns to water, the thing which the change is a change to must not be an affection of this. If it is, the change will be an alteration. Thus suppose a man who knows music ceases to exist, and a man who is ignorant of music comes into being: the man remains the same. Now if knowledge and ignorance of music were not affections of this, it would be a case of coming to be and passing away . . . but as they are, it is a case of alteration' (319^b21–31). That is, if the *terminus ad quem* is parasitic on, called real because of its relation to, anything which remains throughout the change, the change is an alteration; if whatever remains throughout the change is parasitic on the *terminus ad quem*, the change is a coming into existence. This seems a reasonable criterion for distinguishing alteration from coming into existence; if we apply it, however, it is not clear that statues and houses come into existence. It might seem to be rather the case that an arrangement for sheltering is real because it is the arrangement of these bricks, than that bricks are real because they are the bricks of this arrangement. It is notable that Aristotle (though perhaps not for this reason: cf. *De an.* II 412a11–13) wavers on whether artefacts are realities: they have the requisite features (*Met. Z* 1029a27–8) of being particular things which exist 'separately'; yet he sometimes denies that they are realities (*H* 1043a4), and reserves the title of reality for things constituted naturally (1043b21–2).

This is not a serious difficulty, since it is clear that, whether an artefact is a reality or not, that out of which it is made is an underlying thing and a constituent principle. The graver difficulty concerns undisputed realities, like men, dogs, trees. In *Met. H* 1042b25–1043a1 Aristotle puts forward a theory to the effect (see above, p. 55) that for a threshold to exist is for stones to be positioned in a certain way, for ice to exist is for water to be solidified in a certain way, and in general for *F* to exist is for *M* to be differentiated in a certain way, the *M* and the way in which it is differentiated varying with the *F*. It seems natural to take the *F* here as the reality or form which comes into existence, and the *M* as the underlying thing or material factor. Now consider the case of a dog: what is the *M*, for which to be differentiated in a certain way is for a dog to exist? Two answers are possible. We might say the seed: a dog exists if a seed has been fertilized in a certain way. Or we might say flesh and bone: a dog exists if flesh and bone are animated in a certain way, or have a certain kind of life. Aristotle's examples in *Met. H* 2 do not give much guidance, for it is hard to say whether water stands to ice rather as a seed to a dog, or as flesh to a dog. There are passages which suggest that Aristotle would say the former, especially *Top.* IV 127a3–19.

Now it ought to be flesh and bone which is the material factor. It is what, we would say loosely in English, a dog is made of. The matter of *F* is *F* in possibility, and the seed is not a dog in possibility (*Met. Θ* 1049a2) but, if anything, a canine body in possibility (*De an.* II 412b27). It is stuff like flesh which stands to a reality as a reality stands to the *terminus ad quem* of an alteration (*Met. Θ* 1049a27–36—a passage where Aristotle notes that whilst the relation of underlying thing to that which underlies it is the same in both cases, the terms of the relation differ importantly in character). Finally, it seems more correct to say that for a dog to exist is for flesh to have a certain sort of life, than to say that for a dog to exist is for a seed to be fertilized. For a seed to *be* fertilized is rather for a dog to come into existence than for a dog to exist; and a seed may *have been* fertilized, and the dog died.

On the other hand, Aristotle's argument in *Phys.* I will not disclose a factor like flesh and bone. Flesh and bone do not become or turn into a dog; they are not, under any description, a *terminus a quo* of the change to a dog, but rather the *terminus ad quem*. It is the seed which is the *terminus a quo*, the factor which Aristotle's method of handling the matter brings to light, and, incidentally, the factor the disappearance of which makes the change to an animal a coming into existence. (For whilst some features of the seed, e.g. temperature, survive fertilization, they become parasitic on the embryo or embryonic body.)

There is thus a serious gap in Aristotle's argument. Readers have sometimes tried to close it for him by positing something called prime matter, which remains throughout the change but is completely indeterminate. I shall say more about prime matter below, but for the moment we may notice that Aristotle does not say that anything remains, but only that something underlies, in cases of coming into existence, and that according to *De gen. et cor.* I 319^b21–31, cited above, p. 75, if anything did remain in all cases, there would be no such thing as coming into existence, but only alteration. In fact, Aristotle has no need for such a desperate remedy as completely indeterminate matter. He could have recalled that things not only come to be out of, but pass away into, their opposites (188^b3–6) and got at the desired underlying thing in the case of realities by considering not their coming to be, but their passing away: a man passes away into that which is not a man; but he may also be said to be cut up into flesh and bone.

In **190^b10–23** Aristotle draws his conclusions. Everything which comes to be is composite ($^b10$–11); it comes to be out of the underlying thing and the form, and a man who knows music in a way consists of a man and a thing which knows music ($^b20$–2). This may seem abrupt. The most that has been shown is that in all cases we can distinguish an underlying thing and a form, and it is surprising to hear that these constitute what comes into being; what comes into being is surely the form, that to which the change takes place. Aristotle must here have in mind a point developed elsewhere, e.g. *Met. Z* 1033^a28–$^b9$, 1035^a28–34. The proper expression for a thing which is produced or destroyed is an expression like 'a brazen sphere', 'a wooden table'. A sphere, as distinct from a brazen sphere, is what the bronze constitutes, but it is not strictly what the bronze constitutes that the smith manufactures or melts down: he can only make the bronze constitute or cease to constitute it. It is, therefore, the bronze and what it constitutes taken together which can be produced or destroyed. In what way the brazen sphere *consists* of bronze and a sphere Aristotle nowhere says very clearly, but we might say that they are elements in it only in the sophisticated sense that it can be thought and spoken of as either—'logical elements' perhaps. Similarly 'a man who knows music' is a fair expression for the thing which can be regarded either as a man or as an instance of knowledge of music.

Having inked in his account for several lines, (**190^b23–191^a5**), Aristotle says (according to the text as I translate it) that in a way it is not necessary for there to be two opposed principles; one, by its absence and presence, will suffice to effect the change (ᵃ5–7). I suspect these lines are a gloss, taken from 195ᵃ11–14 by a student who did not sufficiently consider that there Aristotle is speaking not of forms but of efficient causes or sources of change; if we accept them as genuine, they present a little difficulty (unless, contrary to what was argued above, we suppose that Aristotle conceived a form as something like house-ness or sphericality). However, Aristotle might have written thus while still conceiving a form as something like a sphere or house (cf. *Met. Z* 1033ᵃ14–15) if he had in mind the mode of exposition to which he refers us in 191ᵇ27–9, one which involves use of the concepts of possibility and actuality. He probably there means the mode employed in *Met. H* 1045ᵃ30–3 (cf. *Λ* 1075ᵇ34–7): when some bronze becomes a sphere, we should not think that a quantity of stuff somehow comes together with a property, sphericality; we should rather think that (through the commonplace agency of a smith) what is a sphere in possibility comes to be a sphere in actuality. On this showing, the factors are a sphere in possibility and a sphere in actuality, and Aristotle could say that one thing, a sphere, by being possible and actual, suffices for the change. And if a sphere which is merely possible is absent, and one which is actual is present, he could say what is said here, that one thing, a sphere, by being absent and present, suffices for the change.

In **191ᵃ7–12** (already touched on above, p. 71) Aristotle says that the underlying nature, i.e. the material factor, must be grasped by analogy: as bronze stands to a statue, and wood to a bed, so the underlying thing stands to a reality or particular thing. This may be understood in two ways. We might take bronze and wood and statues and beds respectively as examples of underlying things and realities: Aristotle will then be saying that 'underlying thing' and 'reality' are just the generic names for things which stand in this relation. Or we might think that statues and beds are not realities, and bronze and wood are not underlying things; but an underlying thing is what stands to something which is a reality as wood stands to a bed. I favour the first interpretation, which seems to me supported by the parallel passages (cited above, pp. 71–2), 195ᵃ16–21 and 1048ᵃ35–ᵇ4 (see also Hesse, op. cit., pp. 336–7). Those who think that Aristotle believed in prime matter favour the second interpretation, and say that prime matter stands to realities as wood to a bed, and that its nature must be grasped by analogy because in itself it is wholly indeterminate. Even if Aristotle believed in prime matter, however, it seems impossible that he is introducing it here. In the first place, according to the more sober, it is only to fire, air, earth, and water that prime matter stands as material factor, and Aristotle would not have called fire or its like 'a reality and a this thing here' (cf. *Met. Z* 1040ᵇ5–10). Second, wood is the proximate 'thing out of which' a bed arises; whatever uncertainty may surround admitted realities like a man or a dog, prime matter is not the proximate thing out of which they arise: that is either seed or flesh. Again, even if Aristotle believed in prime matter, he could hardly have ranked it as a principle without being false to his view (195^b21–3, *Met. H* 1044^b1–3) that we should concentrate on *proximate* causes and principles. Finally, it is incredible that Aristotle should introduce so startling a notion as that of a wholly indeterminate universal substratum in this ambiguous manner, when nothing in the preceding discussion has prepared us for it. We would expect him to say: 'As bronze is to a statue and wood to a bed, so in the case of things constituted naturally there must always be something which remains when they come to be, and this is not different for different things but the same for all. The solution is that it is neither something definite, nor of any definite quality or quantity, etc., etc.' That Aristotle, then, is not here referring to prime matter seems clear; that he does not believe in it at all is argued in the Appendix.

The chapter concludes with a summary of the argument of chapters 5–7: the original theory was that the only principles were the opposites, then that there had to be an additional factor underlying them; it has now been shown what sort of opposites the opposites are, how the principles stand to one another, and what the underlying thing is (**191^a15–19**). In point of fact, Aristotle's account is incomplete: as we have seen, it does not adequately cover a most important group of 'things that are', living things, our concept of which is considerably more complex than our concept of things like brazen spheres. Nevertheless, Aristotle does resolve the difficulties which emerged in chapter 6. These difficulties sprang from conceiving the opposites as indefinite and abstract, like density and rarity or heat and coldness; if we conceive them so, we seem to be forced by linguistic pressure to posit a single kind of matter which is the substratum of everything, and for possession of which the opposites are continually at war. Aristotle shows that language does not in fact oblige us to project this picture on the world. We do indeed speak of opposed factors in connection with all physical things. These factors, however, are not the same in every case; they are different in every case and the same only by analogy. And they are not indefinite opposites, but one is definite. It is also true that in all cases we speak of a third factor. This, however, is again different in different cases, and is not something over and above the opposites, standing to them in a thing–property relationship, but is the same thing as one of the opposites under a different description. This account leaves the scientist, the student of nature in the strict sense, with a world the variety and structure of which is subject to no metaphysical limitations.

CHAPTER 8

The account in chapter 7 was presented as an answer to the problems raised in chapter 6, problems which led physicists to take the first line mentioned in chapter 4, and to say that there is a single universal substratum which undergoes alterations. Aristotle now claims that the same account meets the more fundamental problem which influenced all the 'thinkers of earlier times', and led some of them to monism, the problem of how things can come into and pass out of existence (**191^{a}23–33**).

Aristotle speaks of two ways of dealing with this problem. The second (**191^{b}27–9**) would probably be to say that a thing arises from what is that thing in possibility but not in actuality; see above, p. 78. The 'elsewhere' of ^{b}29 may be, as Untersteiner suggests, the *De philosophia*, but could equally be *Met. Z–Θ*.

The first way (Ross takes the phrase in ^{a}**36** which I translate 'in one way' to mean 'to adopt one way of explaining the matter') is expounded in **191^{a}31–^{b}27** and is difficult. Things come to be, says Aristotle, neither out of what is not nor out of what is, except 'by virtue of concurrence' or incidentally. To establish the first point he uses the doctrine of 190^{b}27, that the lack or opposite is only incidentally that out of which a thing comes. It is awkward, therefore, to illustrate non-incidental coming to be by something dark coming to be pale (^{b}5). His handling of the second point also seems clumsy. He wishes, I think, to say that when A turns into B, though both are things which exist, it is not *qua* thing which exists that A turns into B or that B comes out of A. Apparently, however, he asks us to consider the birth of one animal from another, which is quite a different sort of case of one thing coming from another. And what is it intended to show? The words which I translate 'if a particular sort of animal is to come to be' 'if a particular sort of thing which is' (^{b}23–4) might be more naturally be construed 'if something is to become an animal' 'if something is to become a thing which is'. The trouble with this is that Aristotle would then be saying that a thing can become existent so long as it does not do so from being existent; and it hardly makes sense to talk of something coming to be existent. On my construction he argues that since it is inaccurate to say that dogs come from animals (they do, but strictly they come from dogs), it is inaccurate to say that, e.g., water comes from what is (it does, but strictly it comes from a particular sort of thing which is). On either interpretation the phrase 'for that belongs' (or perhaps 'is present') 'already' (^{b}22–3) is difficult.

Ross's insertion '⟨out of dog or horse⟩' in ^{b}20–1 can be defended on the ground that the MSS. reading 'for instance dog out of horse' is too weird to be correct. This ground seems to me sufficiently strong;

however, if we retain the MSS. reading, and suppose Aristotle prepared to contemplate the possibility of transubstantial change, we can get a smoother sense for the whole passage: 'just as if an animal were transformed into an animal, and an animal of a particular sort into an animal of a particular sort, e.g. a horse into a dog; then the dog would come to be, not only from an animal of a particular sort, but from an animal. It is not, however, as an animal that the dog would come to be, for it is that already (sc. in its equine days). If, then, a thing is to come to be an animal in the strict sense, it must not come to be out of an animal, and therefore if something is to come to be a thing which is, it must not come to be out of a thing which is.'

Aristotle might have done better if, instead of giving a general account illustrated by the case of a doctor, he had considered the kind of specific case which had baffled physicists of the second group (see above, p. 64), the coming into existence of one element like fire or water out of another. In fact, he defers (cf. *Phys.* IV 213^a4–6) discussion of these cases to the *De gen. et cor.* II (v. 329^a27–9); his account in chapter 7, however, is applicable to them. Air, he could say, comes to be out of water in the way in which a plant comes to be out of seed, and its coming into existence is neither a case of replacement—the air does not come from elsewhere or the water go elsewhere (as William III came to England from elsewhere and James II left England for elsewhere)—nor a case of creation, for the air clearly does not come to be out of nothing, but out of the water.

'This nature' in **191^b33–4**, if the text is right, must be the material factor, the underlying nature referred to in 191^a8. Since it has not been mentioned in chapter 8, the phrase is a little surprising, but seems confirmed by the opening sentences of the next chapter.

CHAPTER 9

In this chapter Aristotle contrasts his account of the principles of things with Plato's. He begins (**191^b36–192^a2**) by attributing to Plato two errors. First, Plato accepted uncritically (cf. *Met.* *N* 1089^a1–2 on his 'old-fashioned' approach) Parmenides' dictum that, if a thing comes to be, it must do so either out of what is, or out of what is not. Chapter 8 has shown that this dilemma is improper. Plato, however, grasped the negative horn, and said that things come to be out of what is not. It is not quite clear what Aristotle means by 'what is not' here. As Ross and Cherniss (p. 92) note, in *Met.* *N* 1089^a20–1 he says that by 'what is not' Plato meant the false, and that it is from that and from what is that Plato produces a plurality of things. Now Plato does in the *Sophist* (especially 258–9) equate what is false with what is not, and what is not with what is other than something, as contrasted with what is wholly non-existent. However, I doubt if Aristotle has the *Sophist* theory in mind here. What is not is equated with the great and small in 192a7. Now the phrase 'the great and small' or 'the great–small' is not found in Plato's surviving writings, but in *Phys.* IV 209b13–15 Aristotle says that Plato used different expressions for 'that which participates' in 'the so-called unwritten teachings' and in the *Timaeus.* There can be no doubt that, in Aristotle's opinion at least, 'the great and small' is an alternative expression for what in the *Timaeus* is called 'space' or 'receptacle'; or that in most of this chapter Aristotle is referring to Timaean space (thus in a14 he speaks of a 'mother' element, and in *Tim.* 50 d 3, 51 a 4–5, Plato calls space the mother). I prefer to take it, then, that Timaean space is the only 'thing which is not' which he here has in view. Democritus regarded space as that which is not (*Met. A* 985b5–6, cf. 187a2), and Plato describes it in the *Timaeus* negatively, as not earth, air, fire, or water (51 a 5), so even by the *Sophist* criterion it is something which is not.

Plato's second error was to suppose that if things are the same 'in number', i.e. in fact, they are the same 'in possibility', i.e. in account (**192a2**): he made his pair of principles from which change takes place, the great and the small, the same in account and unreal in the same way, whereas Aristotle makes his pair, the underlying thing and the lack, different. The lack is of itself something which is not (a5), in the obvious sense that the lack is 'that which is not a doctor', 'that which is not a statue', or the like. The underlying thing is a reality in a way (a**6**; similarly 190b25–6), in that a determinate quantity of bronze or flesh is a statue or man (realities without qualification) in possibility (cf. *Met. H* 1042a27–8).

Hence Aristotle's three principles are quite different from Plato's (**192a8–9**). Nevertheless, one of Aristotle's principles, the underlying thing, does the work of Plato's great and small or space: it is a constituent co-responsible with the form for what comes into being (a13–14); and the other, the lack, which Plato overlooked (a12), is responsible for the appearance things have of being indefinite and fluid. This seems to be its evil tendency (**a15**). In these lines I hear ironical echoes of Plato's language in the *Timaeus*: *phantastheiē* (a15), *phantazomenon* (*Tim.* 49 d 1); *oud' einai to parapan* (a16), *mēden to parapan einai* (*Tim.* 52 c 5); *atenizonti tēn dianoian* (a15–16), *chalepon kai amudron, dusalōtotaton* (*Tim.* 49 a 3, 51 b 1, cf. 52 b 3 etc.). As a result of failing to distinguish the underlying thing and the lack, Plato falls into further errors. He makes the material principle, in trying to get into order (cf. *Tim.* 53 b 1), reach out for its opposite, and hence for its destruction (**a19–20**). He makes the material stand to the form as female to male (*Tim.* 50 d 2) and as ugly (disordered) to beautiful (symmetrical); but it is the lack which has these features; the underlying thing has them only by virtue of concurrence

(**192a23–5**; but cf. *Met. A* 988a5, *De gen. an.* I 729a28–30, where Aristotle seems happy to compare the matter–form relationship with the female–male). Finally, Plato makes the underlying thing eternal and indestructible (*Tim.* 52 a 9), which is not true without qualification (**192a25–6** and ff.). Ross prints these lines as if Aristotle had now finished with Plato and were restating his own view of matter; they must, however, be taken closely with what goes before, since the subject of the verbs in a25 must be supplied from the preceding sentences.

Aristotle's explanation of the ways in which the underlying thing is and is not subject to generation and destruction is a little difficult. As 'that in which' it passes away, but as 'that which is possible' it does not (**192a26–7**). 'That which is possible' (as is confirmed by **a30–2**) is the matter or underlying thing as contrasted with the lack; bronze or flesh would be in possibility a statue or man. Ross interprets 'as that in which' as the underlying thing considered as lacking the form which comes to be, 'for', he says, 'when the privation passes away, there is no longer anything "in which the privation is"'. This seems to me awkward and complicated: why did Aristotle not say simply 'as the lack'? 'That in which' is an odd phrase when what is supposed to be 'in' the thing is a privation or lack (the situation is rather that the form is lacked by the matter, than that the lack of the form is in it); and as 'that in which the privation is', the matter would surely pass away rather by virtue of concurrence than of itself (a26).

I suggest the following solution. Plato in the *Timaeus* uses the expression 'that in which' for his space: 50 d 1, cf. 49 e 7. I think that Aristotle is here still using Platonic terminology, and by 'that in which' means simply the lack itself. The parenthesis 'because that which passes away, the lack, is in it' I take to be a gloss by a student who missed the reference to the *Timaeus* and found the passage puzzling. That the passage puzzled early commentators appears from Simplicius ad loc.

That that which is in possibility a statue or the like does not pass away is a point Aristotle insists on elsewhere in similar language, e.g. *Met. Z* 1033a28–b9, *Λ* 1070a2–4, and his reasons seem fair: if you are making an ivory billiard ball, you are not making either the ivory or what the ivory will constitute, a sphere: you are making the ivory constitute a sphere. As the parallel passages show, there is no need to understand 'primary' in **192a31** as 'ultimate' (with Ross): it is better to understand it as 'proximate', as in 193a10.

We may notice that Aristotle's argument establishes only that that which is *X* in possibility is not produced when *X* is produced or destroyed when *X* is destroyed. When *X* is one of the four elements, that which is *X* in possibility may pass away when *X* comes to be, and come to be when *X* passes away. Thus water is air in possibility (*Phys.* IV 213a2–3), and Aristotle would probably say that it ceases to be when it changes into air, and comes to be when air changes back into it. If this is his view, it will not be true without qualification that the material factor is neither brought to be nor destroyed; however, in these chapters Aristotle is discussing changes generally (189b30–3) and taking as his main examples alterations; elements are not mentioned (cf. 184a18–23 and note), and their transformations will not have been to the fore in his mind.

Aristotle's account of Plato's position in this chapter has been severely criticized by Cherniss (pp. 84–6 and elsewhere; see also Ross, p. 566), who holds that Timaean space is a receptacle, not a material substratum, and that Aristotle has simply foisted his own conception of prime matter on to Plato. I argue in the appendix that prime matter is not an Aristotelian conception; and that apart, if we place what Aristotle says about Plato in the *Metaphysics* alongside what Plato himself says in the *Philebus* and *Timaeus*, we get a coherent theory, which is the theory attributed to Plato here, and which, though in some ways attractive, needs emending in just the way Aristotle suggests.

In *Met. A* 988a9–14 Aristotle summarizes Plato's position as follows: 'He clearly uses two causes only, that which is responsible as what a thing is, and that which is responsible as matter. The forms are responsible for all other things as what they are, and the one similarly for the forms. It is also clear what the underlying matter is, of which the forms are said in the case of perceptible things, and the one in the case of the forms: it is a pair, the great and the small.' In *Met. N* 1089b11–14 we are told that the Academy used and spoke of 'large–small, many–few, from which they got numbers, long–short, from which lines, wide–narrow, from which surfaces, and deep–shallow, from which solids'. The use of many–few seems to have been a later development (1087b16–17), and Plato doubtless derived his numbers from large–small. He also held that ideas were numbers (though there were other mathematical numbers): *Met. M* 1080b12–13, 1086a11–12. So according to the *Metaphysics*, Plato held that ideal numbers, i.e. numerical values which determine or express forms, arise out of the one and the great and small, and perceptible objects out of ideal numbers and some or all of these other pairs, the long and short, the wide and narrow, the deep and shallow.

Do we find any echoes of this doctrine in the *Philebus* and *Timaeus*? In *Phil.* 24–7 Plato distinguishes a class of unlimiteds and a class of limits. Unlimiteds are indefinite pairs like hotter–colder, high–low, and, in general, things the expressions for which can be prefixed by expressions like 'more', 'less', 'very', 'slightly', 'excessively' (*Phil.* 24 e 7–8). Limits are things which can be said to be equal, double, etc. (25 a 7–8). Examples might be 'the interval of a fifth' (three quarters the interval of an octave), 'the temperature of boiling brandy' (? half the temperature of boiling water), etc. We are told that a limit combines with an unlimited to constitute a 'mixed thing' or reality (*ousia*) (27 b 9). Now what Plato obviously has in mind here is that, when the strings of a lyre (the low and the high) are in certain numerical ratios, e.g. when one is double the other, consonance results (*Phil.* 26 a 2–4; cf. *Met.* *N* 1092^b14). Consonances are perceptible things, and thus Plato in the *Philebus* is making the principles of perceptible things pairs of opposites and numerical ratios—e.g. underlying thing: the high and low; form: 2:1. Further, the underlying thing here is not high and low sound: it is not that there is twice as much low as high sound in an octave; rather, the underlying thing is high and low strings, and there is twice as much low as high string in an octave. Now the strings are long and short, so strictly the ratio here is in the long and short. If all perceptible things are like consonances (cf. *Met.* *N* 1092^b14–15), and *Phil.* 26 a–b and 27 b suggest Plato thought they were, Aristotle's account seems to be not far wide of the mark.

Similarly in the *Timaeus*. Describing the origin of the world soul, *Tim.* 35 a–36 b, Plato says that a mixture of the slightly mysterious ingredients, being, same, and other, was divided into a number of portions. The subsequent portions stand to the first portion divided off, in ratios which determine a representative section of the diatonic scale; so the passage might be called a deduction of a series of ideal numbers. When he comes to material objects, Plato constructs them ultimately out of two sorts of right-angled triangle, one with sides in the ratio 1, 1, $\sqrt{2}$, the other with sides in the ratio 1, 2, $\sqrt{3}$ (*Tim.* 54 b), and the ratios which concern their numbers, movements, and other powers are all most accurately worked out (though not by Plato) (*Tim.* 56 c, and cf. 68 d 6–8). Space is not mentioned in the discussion of the triangles and the generation of earth, fire, etc., out of them, but we were told it was absolutely necessary (49 a 3, etc.), and, like Aristotle (*De. gen. et cor.* II 329^a13 ff.), we may feel that it is meant to be the substratum to the numerical values determining the triangles: it is what these ratios are ratios in (cf. I. M. Crombie, *Examination of Plato's Doctrines* ii. 222–4). That need not stop it from being what we call space as well. Plato seems to have denied the existence of void (*Tim.* 80 c 3, *De gen. et cor.* I 325^b33), and a person who does that may well identify space with matter or body, as we see from Descartes, *Principles of Philosophy* ii. 11 and 18. And Kant talks of space in a way which (transcendental idealism apart) might appeal to Plato: 'A permanent appearance in space . . . can contain only relations and nothing at all that is absolutely internal, and yet be the primary substratum of all outer perception . . . Something is contained in the intuition which is not to be met with in the mere concept of a thing in general, and this yields the substratum, which could never be known through mere concepts, namely a space which with all that it contains consists solely of relations, formal, or, it may be, also real' (*Critique of Pure Reason* A 284 = B 340).

The *a priori* deduction of ideal numbers we may leave aside; for the rest, Plato's theory is obviously one of those which make the primary features of things those which are quantitative and measurable. It has an affinity with Locke's theory of substance: Locke made the principles of things indeterminate substratum and size and shape, or 'bulk' and 'figure'. And we are in the same camp today, if we try to construct things out of particles of energy or positive and negative changes in certain proportions, saying, e.g., that water is atoms of hydrogen and oxygen in a certain ratio, and that an atom of oxygen differs from an atom of hydrogen in that it comprises a different number of electrons.

In the *De caelo* and *De gen. et cor.* Aristotle criticizes Plato's theory as a scientific speculation. Here he limits himself to the general point that it is wrong to characterize the material element in things as a great and small, or speak of it in terms which are applied according as things exceed or fall short of a norm (cf. 187a16–17). What terms we should use instead he does not say here, but in *Met. N* he tells us that we should characterize it in terms of possibility and actuality: we should call the material element that in possibility which it becomes or constitutes (1089a28–31, b15–16).

This is a philosophical, not a scientific criticism, and I think we may best understand Aristotle's position, if we see Plato as presenting a fact about scientific method as a fact about the world. The scientist breaks down perceptible things into material constituents with quantitative features. What these constituents are which have these features he does not ask and his method cannot reveal; all it can reveal is yet more primitive constituents with features of the same type. For instance, a scientist might discover that a particle consists of two components, a positive and a negative, standing in a certain ratio; but what it is which is positive or negative, is not a question for him at all. Now if philosophers insist on extorting from science answers to questions which are not questions for the scientist, the answers they obtain will be extremely odd. Either it will seem that there are waves with nothing which is waving, energy with nothing which is energetic, triangles with nothing which is triangular, etc., which is a violation of language; or else that there is indeed something which is numbered, measured, and in general has the features noted by scientists, but it is wholly unknowable, so the real stuff of the world becomes an insoluble enigma. Plato's account in the *Timaeus* seems to combine features of both answers: what perceptible things really are is an unknowable substratum which is also nothing at all. This whole way of trying to account for things is as absurd as the procedure of a botanist would be, if he thought that tree-trunks were constructed of obliquely slanting strata or wedges, because these are what they are resolved into by woodmen with axes. Some fine problems in botany might be generated thus: how does the tree form these wedges? What nutritional factors determine this stratification?

The way of speaking recommended in *Met. N* is supposed to protect us from the illusion of an unknowable substratum. It is not true that we have no idea what the basic constituents of matter, whether Platonic triangles, Lockian corpuscles, or particles of energy, in themselves are. They are in possibility what they constitute. You might say that an electron is in possibility an atom, which is in possibility a molecule, which is in possibility flesh, which is in possibility a man, which is not an inscrutable entity, but something of which we can give an extensive and illuminating account. To put it another way, the kind of knowledge we want is the kind of knowledge which can be and is had of things like dogs, beds, and trees; and knowing what these are, it is not only unnecessary but also improper to seek the same kind of knowledge of what they are made of: material constituents are perfectly well understood when we know what they constitute and how, when we know what they are in possibility, and how that which is something in possibility comes to be that thing in actuality.

Whether Plato is in fact guilty, in the *Timaeus* or elsewhere, of confusing science and philosophy, is a question for Platonic exegesis; an advocate for the defence might point out that he is not uncritical of the scientific method of analysing things into their material constituents, cf. *Theaet.* 205–7, though personally I think his handling of Socrates' dream in the *Theaetetus* shows an imperfect grasp of what the limitations of scientific analysis are. But we might notice that the confusion is one to which thinkers are very prone. Thus Eddington begins his book *The Nature of the Physical Universe* with the words 'I have settled down to the task of writing these lectures, and have drawn up my chairs to my two tables'. One of these tables, it appears, is the handsome artefact familiar to him from childhood; the other he calls his scientific table, and it consists of 'numerous electric charges rushing about at great speed'. And he tells us that 'modern physics has by delicate test and remorseless logic assured me that my second scientific table is the only one which is really there'. One would have thought it obvious that the remorseless logic of modern (herein unlike Presocratic) physics does not permit its votaries to distinguish different kinds of tables, much less inform them which is really there; but the same absurdity is found in Russell. In *Problems of Philosophy*, chapter 3, we read: 'What is the nature of this real table, which persists independently of my perception of it? To this question physical science gives an answer, somewhat incomplete it is true, but yet deserving of respect so far as it goes.' Physical science gives no answer whatever to Russell's question, because it is a philosophical, and not a scientific, question.

BOOK II

CHAPTER 1

In this chapter Aristotle first (**192^{b}8–193^{a}9**) introduces the notion of nature, and then raises the main question of the book, whether it is only the matter of a natural object, or its form too, which we can call its nature.

The notion of nature is introduced by means of a distinction, which appears in ordinary speech (192^{b}11–12), between natural objects and non-natural objects. We set animals, plants, and things like earth, air, fire, and water, apart from things like beds and coats, and regard the former as due to nature, or natural, in a way in which the latter are not (^{b}**8–16**). It might seem from Aristotle's examples that the class of non-natural objects is the same as the class of artefacts, but Aristotle speaks of 'other causes' in the plural (^{b}8–9), and would probably have included things which come to be by chance or 'automatically' (for which see *Met. Z.* 9).

It is hard, I think, to deny that we do draw this distinction. Aristotle next inquires into its basis (^{b}**13–20**): on what principle do we classify dogs as natural objects but not beds? His answer is that a natural object has in itself a source of its changing and staying put, whilst a thing like a bed has not. And this being the difference between natural and non-natural objects, he is able to define nature as an internal source of, or factor responsible for, a thing's changing and staying put (^{b}**21–2**). It is hardly necessary to point out that Aristotle conceives the word 'nature' as applying, not to some single all-pervading demiurgic force, but to that factor in a thing which we call its nature; so that there is for him no such thing as nature over and above the nature of this, the nature of that, etc. Elsewhere (*Met. Λ* 1070^{a}12) he says that nature is 'a kind of disposition'.

What does Aristotle mean by an internal source of change, and is he right in thinking that it is what differentiates natural objects? Change here includes not only movement, but change of every sort (^{b}14–15) such as, no doubt, the formation of organic parts like leaves and teeth, and what Aristotle has in mind at this stage is, I think, a very simple point. If we are asked why a stone when released (not thrown) from on high, falls to the ground (one kind of change of place), we may reply, simply, 'Because it's a stone'. If we are asked why a dog when it sees a rabbit gives chase (another kind of change of place), we may reply 'Because it's a dog'. If we are asked why this tree puts out broad flat leaves in spring and keeps them through the summer, we may reply,

'Because it's a beech'. In these cases, rightly or wrongly, we do not feel it necessary to look outside the thing, to account for its behaviour. And wherever we feel that we can explain a thing's behaviour, partly at least, without looking beyond the thing, we think that its behaviour, and the feature it acquires or retains, is natural. It is natural for stones to fall, natural for dogs to chase rabbits, it is the nature of beeches to have broad flat leaves.

Has Aristotle here a criterion for distinguishing natural from non-natural objects? In the examples above, the answer 'Because it is an *f*' is given to questions of the form 'Why does it ϕ in circumstances *C*?' The answer is valid, if ϕing in *C* is definitive of *f*s, if an *f* is identified as a thing of the sort to ϕ in *C*. Now Aristotle's statement that artefacts do not have in themselves the source of their making (b**28–9**) suggests that he hopes to distinguish processes of manufacture from processes of growth. On our present showing alone, however, it is hard to see how we can. For if it is asked why iron when heated, hammered, filed, etc., comes to be a saw, we may surely reply 'because it is iron'. Nor would it help Aristotle to concentrate on the behaviour of completed artefacts and mature living things, for if a dog gives chase when it sees a rabbit because it is a dog, why not say that my washing machine washes and spin-dries my shirts when I press the programme-button *C* because it is a washing machine, i.e. has the internal structure and disposition of parts of a washing machine?

To distinguish natural things, then, as things with an internal source of change, requires further preliminaries. In the first place, it will be necessary to distinguish the kind of explanation we give when we say 'It became a saw because it was acted on in such and such a way'—a kind of explanation which can also be given of the growth of living things—from the kind of explanation we give when we say 'It became a saw through the skill of the smith'. And second, it will be necessary to show that natural processes occur through something analogous to the skill of the artisan, but internal to the things undergoing them, and internal not just by chance, as knowledge of medicine is internal to the doctor who treats himself (b**23–7, 30–2**), but as a matter of definition. Aristotle is not unaware that he has these tasks ahead of him, but sees them in terms of the matter–form distinction. The kinds of explanation to be distinguished are that in which the change is referred to the matter as source and that in which it is referred to the form, and what has to be established for a natural process is that it can be referred to the form of the thing undergoing it. Since Aristotle tends to see the natural movements of earth, fire, etc., as changes due to the matter of things, the naturalness of inanimate natural things and their behaviour does not get established; but he would probably have said that it did not need establishing, and that the controversial issues concern living things.

In the second half of the chapter, then, **193^a9–$^b21$**, Aristotle opens the question whether of the two internal factors distinguished in *Phys.* I, matter and form, both or only one can be the nature of a thing, the internal source of its behaviour. Often it is the matter, what the thing is made of, which is the source of behaviour to be explained. When we say that the stone falls because it is a stone, we mean that it falls because it is made of stone (i.e. for Aristotle, a kind of earth). In 192^b19–20 Aristotle noted that an artefact like a bed has an internal source of behaviour, though not *qua* bed but *qua* wooden; this will be a material source: it is natural for a bed to catch fire inasmuch as it is made of wood. In **193^a9–28** Aristotle presents the case of those who hold that the source of natural behaviour is always and only the matter. If they are right, then that living things grow with the features they do can be explained by and only by the action of external things on their matter in accordance with the nature of their matter. Aristotle argues against this possibility formally in chapter 8, but here offers some general, what he might call logical, considerations, which suggest that the form of a thing too may have a claim to the title of nature.

First, the word 'nature' and its cognates is used in the same manner as the word 'art' (**193^a31–3**). We speak of art, not when something could be made into a bed or the like, but when it actually is a bed, has the form of a bed; hence we should say that a thing has a nature, not when it could become flesh or bone or the like, but when it actually is flesh, has the form we give when we say what flesh is (**$^a33$–$^b3$**. Or, perhaps better, a thing has a nature, not under the description, e.g. 'earth' 'fire', under which it is flesh in possibility, but under the description under which it has the form we give, etc.). We might object that what is flesh in possibility, sc. earth or fire, is still something natural; but Aristotle is claiming that there is a use of the word 'natural' parallel to that of 'artificial', and earth or fire would not be natural in this sense (except insofar as it is a *terminus ad quem* of a natural change, and hence a sort of form). As for the idea of artefacts having art in them or being art, we do talk in this way—'There's art for you', we say. 'A lot of art's gone into that'—and according to Aristotle we are not wrong to do so. Art, he holds (e.g. *De part. an.* I 640^a31–2), is the account of, or prescription for, the work of art, without the matter. That is, art, like nature, is always the art of something definite, the art of making a table or restoring men to health or the like, and is, in fact, the form which the artist has in mind, or intends, for the material, the pieces of wood or the patient's body. While he has it in mind only, it is only a possible form; it is realized in the material when the work is finished, and thus actually exists only as what the material constitutes. (Cf. *Met. Z* 1032^b5–14.)

Further, if the form of a thing is its nature, it has a better claim to be called its nature than the matter, since an actual *x* has a better claim to be called an *x* than a possible one (**193b6–8**). By the matter here Aristotle again probably means less (e.g.) the seed in the case of a man than the flesh and bone: if what this constitutes is the nature, then being what it constitutes only in possibility, it is the nature only in possibility (cf. the similar point about the term 'reality' in *Met. H* 1042b9–10).

Again, Antipho's argument of 193a12–17 that the matter must be the nature because beds give birth to wood, not beds, tells equally on the other side. If the nature of a thing is that element in it which is like what it gives birth to, the nature of a man will be a man, i.e. what the flesh and bone constitute (**193b8–12**).

Finally, (**193b12–18**) Aristotle offers an obscure argument based on the Greek word for nature, *phusis*. He might be taking it as a possible word for birth; it is so used by Empedocles, DK 31 B 8. In that case his point is that *phusis* in the sense in which it is used for a process, i.e. in the sense of birth, is *phusis* of the form, e.g. a man, not of the matter, e.g. menses. Alternatively, as most commentators suppose, he is making play with the fact that *phusis* comes from a verb which in the passive means 'to be born' or 'to grow' (cf. the Latin *natura*). Suggesting, then, that *phusis* might be used for a process, sc. growth (or perhaps simply—the text is ambiguous—for coming to be), he says that nature ought to be what this process is a process towards, not what it is a process from, and what it is a process towards is the form. Exactly why the process should not proceed from nature, as doctoring proceeds from knowledge of medicine, is unclear; however, in *Phys.* V 224b7–8 Aristotle says that changes are named after what they are changes to, rather what they are changes from, and it is certainly true that *a* growth is what growth is into, not what growth is out of (if a man does not shave, hair grows out of his chin into a beard, and we call the beard a growth, not the chin). Again, we rather say that seeds are seeds of what they grow into (cf. *De part. an.* I 641b33–6) than that trees are trees of what they grow out of.

Wieland has emphasized that according to Aristotle the nature of a thing is only *a* source of its behaviour, not *the* source. This is perhaps the right time to consider just how far Aristotle thinks that internal sources are responsible for behaviour.

It is a central thesis of *Phys.* VIII that nothing changes itself, that whatever is subject to change is changed by something else. Aristotle argues this separately for kinds of stuff like earth and fire and for living things; this is not only because he thinks that different accounts are needed for the downward movements of stones, etc., and the appetitive movements of animals, but also because he thinks that the former originate from matter and the latter from form, and matter and form are in his opinion sources of change in different ways.

A stone (piece of earth) is heavy, and heaviness stands to earth as dispositional knowledge to a man (255^{a}34–^{b}6). Just as, when a man who knows dispositionally what a dog is encounters a dog, he knows actually that it is a dog, since that is what it is to have dispositional knowledge, so a heavy object, unless anything stops it, goes to and stays at the centre of the earth, and that is what it is to be heavy (255^{b}15–16). Now as something, e.g. a teacher, makes a man know something dispositionally, so something makes earth or stone come into existence (perhaps by effecting a qualitative change in fire or water, cf. *De gen et cor.* II 331^{a}32–3). Whatever does this is in a way responsible for the stone's movement. And anything removing an impediment, e.g. cutting a string by which the stone is suspended, is also in a way responsible for the movement (256^{a}1–2). But beyond this, though the point is not brought out, there is nothing responsible for the stone's movement, because just as a man who passes from having knowledge but not exercising it to exercising it does not therein undergo change (*De an.* II 417^{b}1–16) so a stone which passes from being heavy but impeded from moving to moving towards the centre of the earth does not therein undergo change.

This is a bold account, but whether or not a man who exercises dispositional knowledge is therein changed, it is clear that a stone which moves does undergo change, viz. change of place. It seems to me that Aristotle must say, either that it is moved by the centre of the earth, or that it moves itself. I suspect that, despite his general protestations, he would say the latter. He several times uses a word for nature which seems to mean active striving: *hormē*, *An. po.* II 95^{a}1, *Phys.* II 192^{b}18–20, *Met.* Δ 1023^{a}9, 18, 23, and, most important because it is a careful passage, *E. E.* II 1224^{a}18–^{b}9. In every case but the last the nature involved seems to be the material element in a thing, and the last is not a serious exception, because Aristotle is there explaining freedom and constraint in human action by comparison with natural and constrained movements on the part of things like stones. This strongly suggests that he thinks that the material of a thing can be a source of change because it has an active tendency to change independent of any external cause. (This point is interestingly brought out by A. D. P. Mourelatos, 'Aristotle's "powers" and modern empiricism', *Ratio* 1967; Mourelatos, however, does not, as I do, associate the active tendency specifically with the material factor.)

Turning to living things, Aristotle admits that animals may correctly be said to move themselves from place to place. 'However, there is no reason why it should not be the case, and perhaps it is necessarily the case, that many changes are produced by the environment in the body, and some of these change the thought or the appetition, and that changes the whole animal' (253^{a}15–18). Again, 'Animals are responsible for only one sort of change in themselves' (sc. change of place) 'and even for that they are not properly (*kuriōs*) responsible. The cause is not from the animal itself, but there are other natural changes present in animals, which they do not undergo through themselves, e.g. growth, decay, breathing, which they undergo when staying put and not changing place. The cause of this is the environment, and many things which come in, for instance, in some cases, food' (259b6–12). In *De motu* 10 Aristotle says that the movement of animals originates in some obscure stuff called connatural breath (*sumphuton pneuma*), but how directly the changes in this breath are the result of alterations in sense-organs or of 'things coming in from the environment' and how, if at all, they are determined, in the case of a man, by his intentions and general view of what is the best thing for man—this is left in pitch darkness.

In these passages Aristotle seems to be saying, not just what he says elsewhere, e.g. *De an.* III 433b10–12, that animals and men are moved by objects of awareness, but that they are moved mechanically by objects affecting their sense-organs. It is hard, however, to believe that this is really his view. Perhaps what he would have wished to say is something like this. Any action by a man or animal must admit of a description, e.g. 'movement of ten stone of flesh and bone two yards to the north' under which it is proper to demand a mechanical explanation of it; and it is always possible to find something which does explain it mechanically. It seems possible to hold this while denying that an action, which under another description we would call intentional or a movement of pursuit or avoidance, can be explained mechanically by the action of some external thing on the body of the thing which performs it.

CHAPTER 2

Aristotle here discusses the scope of the scientific study of nature, considers how the student of nature in the sense of the natural scientist ought to proceed. The chapter also contains a fresh approach to the question whether the form of a thing can be called its nature: if, as Aristotle maintains, the student of nature should consider both the form and the matter in natural things, that is an indication that the title of nature belongs to both.

In the first part of the chapter, **193b22–194a12**, Aristotle contrasts the student of nature with the mathematician, and in particular the geometer. The geometer is concerned with shapes, which are things which supervene on natural objects (**193b27**). By 'things which supervene' here Aristotle probably means not just 'accidents', things which are affections of natural things, but features about which it is the business of the student of nature to argue and attempt demonstrations: cf. *De an.* I 402a15, 402b16–403a1, *Met.* Δ 1025a30–2. It does not follow, because mathematics is concerned with shapes, either that the student of nature is wrong to discuss the shapes of natural objects like the sun and the moon (**193b27–30**) or that there is no difference between mathematics and the study of nature. Though the shapes the mathematician studies are things which supervene, he does not study them as such (**b32–3**), that is, he does not (e.g.) try to prove that spherical shape is the shape of the earth (for the student of nature's proof of this, v. *De caelo* II. 14); instead, he separates shapes, as those who posit ideas (the Academy) separate physical things (**193b33–194a1**).

What does Aristotle mean here by separation? Plato's original theory that there are ideas separate from the things which partake in them (Plato, *Parm.* 130 b 2–3) seems to have been simply that for some values of x, there is a thing which is x over and above, i.e. not identifiable with any of, the perceptible or physical things which are x. Thus (following N. R. Murphy's interpretation of a slightly controversial passage, *Interpretation of Plato's Republic*, p. 111 n.) there is something equal which is never unequal, and which is therefore separate from all perceptible things, any one of which, though perhaps equal to something, will be unequal to something else (*Phaedo* 74 b–c). Aristotle sometimes criticizes the Academy for making natural things such as a man, a horse, separate in this way (e.g. *Met. Z* 1040b32–4); however, as Ross observes, this should not be what he has in mind here, since geometers do not separate shapes in this way. Rather, they separate them in thought or account.

Aristotelian forms are, up to a point at least, separable in account (*Met. H* 1042a28–9), that is, an account can be given of the form of a thing which is separate from, does not involve, the account of its matter. Thus if some bronze constitutes a sphere, we can give an account of a sphere without mentioning bronze. It should be observed that though, in consequence of this, we might say that sphericality is separate from bronze, or from what it would be to be bronze, we are not entitled to say that a sphere is separate from bronze. Spheres are not something over and above lumps of material.

Aristotle considers that geometers take advantage of this separability of shapes in account, though for reasons to which we shall come shortly he speaks of it as separability from change (193b34, 194a5), not as separability from matter. His formal account of the geometrical approach is in *Met. M* 3. Perceptible things are subject to change (or, perhaps, are movable), and we can consider what is true of them purely as subject to change without committing ourselves to the view that there is anything subject to change over and above perceptible things. In the same way we can consider what is true of things which are subject to change purely in so far as they are solids or planes or lengths (1077b28–9; i.e., perhaps, in so far as they extend in three dimensions, two, or one), without supposing that there are any such entities over and above things subject to change. This highly realist account of geometry seems to be in Aristotle's mind here, since he says that the geometer considers natural lines, though not as natural (194^a10–11). We might notice, however, that there is a slight discrepancy between our passage and *Met. M* 3: there he does not make optics and harmonics the converse of geometry as he does here in **194^a9–12**, but says that they treat sight and sound as lines and numbers (1078^a15–16). This discrepancy suggests to me that *Phys.* II. 2 was revised after *Met. M* 3 had reached its present form.

If this is what separating is, why do the geometers get away with it, and those who talk about ideas not? To see the answer to this we shall have to go into the doctrine, prominent in this chapter and throughout Aristotle, that 'all natural things are said like snub, e.g. nose, eye, face, flesh, bone, animal as a whole, leaf, root, bark, plant as a whole, for the account of none of these things is without change, but always involves matter' (*Met. E* 1025^b34–1026^a3).

How is snub 'said'? The Greek word translated 'snub', *simos*, means strictly 'snubnosed' (cf. Plato, *Theaet.* 209 c 1) and hence applies strictly to men, and to noses, as in **194^a6**, only by a solecism: v. *Met. Z* 1030^b32–4. (Similarly 'squint' belongs to men rather than eyes, 'brachycephalous' to men rather than heads.) This being so, a thing is called snubnosed, not because there is a definite feature, snubnosedness, in it, but because there is something else, concavity, in something else, its nose. 'Snubnosed', then, is something said because something is in something else (*Met. Z* 1037^b3–4), because 'this is in this' (1030^b18, 1036^b23–4, *De an.* III 429^b14). It is otherwise with expressions like 'concave' (*Met. E* 1025^b33–4, *Z* 1037^b2) and 'white'. A thing is called concave or white because there is a definite feature, concavity or whiteness, which belongs to it. It is possible, then, to separate concave, white, spherical, etc., in a way in which we cannot separate snubnosed, squinting, and the like. An account of snubnosedness must bring in both concavity and noses, an account of squinting both intersection and eyes.

The doctrine that expressions for natural things are all like 'snubnosed' and not like 'concave' is, I think, defensible, and has consequences for Aristotle's conception of a reality. On the one hand, we can now see why expressions like 'a man', 'a dog', are expressions for realities in Aristotle's sense: a man is not the man *of* anything further, because we speak of a man only when something is in something else, when certain capacities are in flesh and bone or a particular sort of organic body (cf. *Met. Z* 1033^b24–5, 1034^a6–7, 'Callias or Socrates is like this bronze sphere, . . . is this sort of form in this flesh and bone'). The point is clearest with an artefact, like a penny. A penny need not be the penny of anything further, because by a penny we mean a cylinder of copper.

On the other hand, Aristotle may have to modify his opinion (see above, pp. 70, 72) that the form of a thing has the best claim to the title of reality. If 'a man' is like 'snubnosed' it will not be a wholly correct expression for a form in the sense of that which the matter of a thing constitutes, and in its place we shall have to substitute expressions for affections, dispositions, abilities. This too can be seen from the example of the penny: if a penny is a cylinder of copper or in copper, what is constituted is a cylinder, not a penny; it is incoherent to say that copper constitutes a cylinder of copper (cf. 1029^b18–19), and for anything else to constitute a cylinder of copper is not one thing but two, since it is one thing to be copper and another to be a cylinder (compare above, p. 74).

However this may be, the comment on 'those who talk about ideas' is that they treat expressions like 'a man', 'an eye' as if they were like 'a sphere', 'concave' when in fact they are like 'snubnosed'. This seems to be a fair comment on Plato's later thought, if, as suggested in the commentary on *Phys.* I. 9, Plato held that natural things are assemblages of triangles in space, and the perfect account of them would be one specifying the ratios determining them.

Why Aristotle talks of separating things from change, instead of from matter, is not quite clear, but two reasons may be suggested. First, as it is presented in *Phys.* I, the study of nature is primarily the study of things subject to change. It is the study of things with matter, only because, according to Aristotle, logical analysis reveals that a material factor is presupposed in any case of change. Influenced by this consideration he might think that we discover that expressions such as 'a man', 'a dog' are like 'snubnosed' when we consider how they come to be and pass away (cf. *De an.* II 414^a22–7). Second, Aristotle does not think that the objects of geometry are considered altogether apart from matter. Even an expression like 'straight line' is like 'snubnosed' in a way: we have a straight line when we have duality in a continuum, or something like that: *De an.* III 429^a18–20. Aristotle seems here to be scenting the possibility of a pure abstract geometry in which shapes are defined arithmetically or algebraically; ordinary geometry is then the study of these formulae in two- or three-dimensional space.

The tendency of the Academy was to confine the study of nature to the study of the forms of natural things. In the second part of the chapter, **194^a12–$^b15$**, Aristotle opposes the contrary error, of confining it to the study of their matter. Again he uses the example of snubnosedness ($^a13$). If asked to develop the analogy, he might say that to explain what snubnosedness is, we have to bring in noses, but we cannot explain it simply in accordance with the principles of noses. The principles of noses are (?) the principles of respiration, which are powerless to illuminate snubnosedness except in so far as it involves, say, a deviated septum. Anyhow, to explain natural things like plants, animals, blood, we have to bring in what they are made of, but we cannot explain them by, or exclusively by, the principles of what they are made of, fire, earth, etc. The principles of fire and earth, that the former goes up and is hot, and the latter goes down and is cold, will not suffice to account for the natural things of which they are the matter.

This, as is well known, is the view Aristotle puts forward elsewhere, most emphatically in *De part. an.* I. 1. The student of nature must consider not only what things are made of, but also what they are for. It is worth emphasizing that there as here Aristotle insists on the importance of considering *both* aspects (e.g. 642a14–15). He nowhere says, but, as we have seen, denies, that the student of nature should concentrate exclusively on the form.

In **194a21–b9** there are some arguments against ignoring the form. First (**a21–7**), medicine, a practical science, deals with both health and that which stands to it as underlying thing, so the study of nature should do likewise, deal with both matter and form. Second (**a27–33**), if you study that which is for something, you should study what it is for; the matter in a natural object is for something, and the form is what it is for, since it is a natural *terminus ad quem*, so the student of nature should study both. We cannot assess this argument until we have seen how Aristotle connects the notions of form, end, and thing for the sake of which, but the two senses of 'that for which' mentioned below in **a35** are 'that to obtain which' ('He did it for money') and the beneficiary ('He did it for his aunt'). Third (**a33–b8**), there are two sorts of knowledge we can have of a thing's matter, one concerning its manufacture, the other concerning its use. The former does not involve any special knowledge of the form, but the latter does. Now we must have both sorts of knowledge of the matter of artefacts, since their matter is either manufactured (like bronze) or at least rendered easy to work with (like wood which has to be seasoned); but the matter of a living thing is not manufactured, so the chief question is: how does the living thing use it?—a question to answer which we need knowledge of the form. The distinction between knowing how to make and knowing how to use is acceptable (and seems to be taken, complete with example, from Plato, *Crat.* 390), but its application to the study of natural things seems to me awkward and obscure. Finally, (**b8–9**) matter is relative, and varies with the form. That is, matter is the matter of something or matter for something—this wood is the wood of this tree or the material for a chest of drawers—and we consider matter in relation to form rather than form in relation to matter; we ask what would be suitable material for a coat or an aeroplane, not what would be a suitable form for wool or aluminium.

The student of nature, then, has reason to consider the form. He should do so, however, only to the extent of considering what things are for (**194b11–12**), that is, he should be satisfied to discover a natural thing's function, what it characteristically does (see below, pp. 101–3), and how its parts enable it to do it. Thus he should be satisfied with an account of the form of an eye such as 'An eye is an organ which enables us to perceive colours'. And 'he should confine himself to things which are separable in form, but are in matter'. By 'in form' Aristotle probably means 'in account' (cf. 190a16), and his point is probably that, for the purposes of the study of nature, we should take a form, what the matter constitutes, to be something like a man, a plant, an eye, things which are strictly (see above, p. 95) 'forms in matter'. Similarly in 192b1 he speaks of 'natural forms which can pass away', a description applicable to entities like a man, a leaf. That which is separable (b14) is probably that which is separated in account, e.g. roundness, what it would be to be to be a man (if that is separable); and it is for first philosophy, for discussions like those in the *Metaphysics*, to determine how it is with such things, i.e. whether such expressions apply to entities in fact separate from quantities of matter, or have application only because it is possible to give separate accounts, distinguish separate possibilities, etc.

CHAPTER 3

This chapter is concerned with the various sorts of thing which may be given as *aitia*. It consists of two enumerations, one of senses in which a thing may be said to be an *aition*, the other of ways (*tropoi*) in which we may meet a request for an *aition* in any one of these senses. *Aition* is traditionally translated 'cause', and I follow that practice, but we should be careful not to be misled by it. We talk of causes operating, and producing effects. Aristotle has no such expressions, and when he wishes to talk of things to which *aitia* stand as *aitia*, he does so in precisely those terms (e.g. 195b7). The Greek word *aition* (connected with the verb 'to blame' 'hold accountable') is used considerably more widely than the English 'cause'. *X* is called an *aition* in respect of *Y*, if it is responsible for *Y* in any way whatever, if *Y* can for any reason be set down or ascribed to it.

This being so, we should not expect from Aristotle's discussion of 'causes' light on those problems about causal efficacy and causal connection which were bequeathed by the British Empiricists. Aristotle does not seem to have been conscious of any difficulty in the notion of an agent exerting power and effecting a change, and in so far as he was conscious of difficulties about the connection between an earlier and a later event, he thought of them as difficulties rather about continuity than about causality: see *An. po.* II. 95a22–b12, where for a detailed discussion of how the past joins on to the future he refers us to his treatise on change, i.e. (probably) *Phys.* V–VIII (*Phys.* VIII. 8 with its 'appropriate' arguments against Zeno seems particularly to the point). The discussion of *aitia*, on the other hand, is rather a discussion of explanation, and the doctrine of the 'four causes' is an attempt to distinguish and classify different kinds of explanation, different explanatory roles a factor can play.

Aristotle holds, as is well known, that these can be reduced to four (195^{a}15): one thing can be responsible for another in that it stands to it as matter, form, source of change, or end. Before we come to details, a couple of preliminary points.

'We do not know', says Ross, p. 37, 'how Aristotle arrived at the doctrine of the four causes: where we find the doctrine in him, we find it not argued for but presented as self-evident.' Although the fourfold classification of the contents of the universe in Plato's *Philebus* (23 c–d) may be a forerunner of Aristotle's classification, and although, as we shall see (pp. 118–20), Aristotle wishes to connect the kinds of cause with the kinds of middle term used in syllogisms, it is obvious (cf. Wieland, p. 262) that the doctrine with which we are presented here is the immediate result of a survey of how we ordinarily speak: see 194^{b}24, 34–5, 195^{a}3–4, 15. Aristotle goes through a lot of things which people in practice hold responsible for other things or (198^{a}14–20; see below, p. 111) give in answer to the question 'On account of what?' and finds that they fall into four groups. A further kind of cause is not ruled out *a priori*, but cannot in fact be found (*Met. A* 983^{b}5–6, 988^{b}18).

As to the purpose of the doctrine, whilst it is of philosophical interest in its own right, Aristotle presents it as an aid to the natural scientist. He will be able to refer particular objects of inquiry to one or another of the four main types of cause (**194^{b}22–3**), and when faced with something of which an account needs to be given, he should look for explanatory factors of all four types (198^{a}22–^{b}9; see below, pp. 113–4. We might say that the doctrine is intended to have both heuristic and therapeutic value: heuristic in an obvious way, therapeutic in that it removes difficulties which might confuse us if we thought that all explanatory factors must explain in the same way. Thus, suppose one man thought that there are plants because there are leaves, roots, and stems; another that there are leaves, roots, and stems because there are plants; a third that there are plants because there are seeds. Aristotle would say that there is no real dispute here, since each party is bringing forward a factor which is explanatory in a different way. The leaves, root, and stem account for the plant in the way in which bronze accounts for a statue, the plant accounts for the leaves, roots, and stem in the way in which health accounts for surgical instruments, and the explanatory role of the seed is in some ways like that of the bronze in the case of the statue, and in some ways like that of the sculptor. We may think that the natural scientist is not in practice likely to need much philosophical assistance of this sort; but philosophers have themselves, perhaps, been misled at times by a feeling that every explanatory factor should be responsible for what is to be explained in the same way—usually in the way in which a murderer is responsible for his victim's death.

So much on preliminaries; we may now look at the types of cause in detail. The material cause is characterized in **$195^{a}19$** as 'that out of which'. It is what the thing to be explained is made of or can be cut up into. We might notice that something is matter only relative to something. Thus wood may be matter relative to a chest, but form relative to earth (cf. *Met. Θ* $1049^{a}19$–24). It is a little surprising to find premisses listed as material cause relative to conclusions (**$195^{a}18$–19**), but we may recall that Aristotle regarded his syllogistic as analysis, and think of the premisses of a syllogism as an analysis of the conclusion. Thus take the second figure (*An. pr.* I $26^{b}34$–5 and ff.). Conclusions in the second figure are to the effect that it is false that *N* belongs to *O*, and this is false when there is something *M* which belongs to one and not to the other. In *An. po.* II. 11, the other main source besides our chapter for the fourfold classification of causes, instead of a material cause, Aristotle speaks of things which are such that 'if they are, this necessarily is' ($94^{a}21$–2, cf. 28). He is probably not thinking there simply of premisses relative to a conclusion, but of another aspect of the material cause to which we shall come below, p. 116.

The formal cause is introduced in a manner calculated to please the Academy, as the paradigm or model (**$194^{b}26$**), and with a Platonic example, the ratio of two to one (**$^{b}28$**). In **$195^{a}16$–21**, however, he says that syllables stand to their letters, artefacts to the stuff of which they are made, bodies (i.e. living bodies) to fire, etc., wholes to their parts, and conclusions to premisses, as their formal cause, and here the formal cause is clearly that which the matter proximately constitutes.

Aristotle habitually uses two slightly puzzling expressions for the form of a thing. One is *to ti ēn einai*, on which see above, pp. 58, 72–3; the other (see, for instance, $194^{b}27$, 29) is *logos*, an account or formula. It might be thought that this is short for 'that of which the account is an account', 'that which the account is of', a phrase which is certainly found, e.g. $191^{a}13$, *Met. Z* $1035^{a}21$, cf. $193^{a}31$. However, Aristotle so often calls the form simply the *logos* (e.g. *De an.* II $412^{b}16$, *Met. Z* $1035^{b}29$, $1039^{b}20$, 24, *H* $1042^{a}28$, $1044^{b}12$), that it is hard to escape the conclusion that he really thought that a form is an account, or at least that it stands to that of which it is the form, as an account to that of which it is the account. It may seem very odd to say that the form of a man is the account of a man, especially if we think of an account as a descriptive speech. But that which is given or expressed in a speech may be called an account, as well as the expression of it, and a prescription or formula as well as a description may be given in a speech. (In cookery books we find prescriptive, not descriptive, accounts of stews, cakes, etc.) When Aristotle speaks of a form, e.g. health or the form of a man, as an account, I think he is considering the form as possible (cf. p. 60), thinking of it as that possibility the realization of which would be my health or me. Such a possibility might be given as the prescription for health, the formula for a man.

The third sort of thing which can be called a cause is a source of change or of staying unchanged. Aristotle's examples are heterogeneous, and some closer to the thing to be explained than others. Art is closer to the statue than is the artist (**195b23–4**), and the seed (**a21**) is presumably closer than the father (**194b30–1**) to the child (cf. *Met. H* 1044a35). Strength is said (**195a9–11**) to be the source of change relative to hard work. Presumably, then, moral states, virtues and vices, would be sources of change relative to voluntary actions, and closer sources than the deliberate agent (**194b30**). That being so, it is misleading to call Aristotle's sources of change efficient causes (we would not call injustice the efficient cause of a murder) and wrong to think of them as Humean causes. We may notice, in contrasting Aristotle with Hume, that Aristotle says that actual causes are always contemporaneous with the things for which they are responsible, not antecedent (**195b17–18**; the point is developed in *An. Po.* II 95a22–b1, cf. also 98a35–b4).

Finally, there is the end. This, we are told, is what the other things are for, and the best thing (**195a24–5**). 'The other things' sounds a vague phrase, but may be taken fairly literally. Not only are the organic parts and natural behaviour of living things, according to Aristotle, for something, but also dispositions, like strength and medical knowledge, are for things, for hard work (195a10) or health (cf. *E.N.* I 1094a1–9).

That arts and artefacts have objectives which are in some sense goods may be generally agreed; but Aristotle notoriously hopes to find, outside the sphere of rational action, factors which are ends in the sense of being what is best and what other things are for: can this be seen otherwise than as a foolish mistake?

In 194a29–30 Aristotle suggests that wherever we have a continuous process of change, its *terminus ad quem* is what it is for. This idea is found in Plato, e.g. *Phil.* 54: the process of shipbuilding is for the ships in which it terminates rather than the other way round. However, it is not much use, first because even if a process is rather for its end than the end for the process, still it might be that neither is for the other, and second because, as Aristotle himself observes (194a30–3), processes may end in bad states such as death or disease; we cannot establish what a process is for just by observing where it ends, but must know independently what 'the best' end for it is.

Besides ends of processes, however, Aristotle speaks of ends of physical objects, and the end of a physical object is its work or function (*Met.* *Θ* 1050^{a}21, *E.E.* II 1219^{a}8); can he establish that things other than artefacts have functions, and then argue that performing its function is what is best for a thing and what its parts, dispositions, etc., are for?

Plato defines the function of a thing as that which it alone or it better than anything else can do (*Rep.* I 353^{a}10–11). This definition as it stands will not do for Aristotle, unless he is prepared to allow that the best thing for men might be garrotting widows or detonating thermonuclear bombs. R. Sorabji (*Philosophical Quarterly* 1964, 291 ff.) suggests that the discharge of a function must confer some good, but this idea is foreign to Plato and Aristotle, would not stop some trivial occupation (e.g. comparing the incidence of hiatus in various works of Plato) from being the function of man, and would stop the attempt to define function independently of goodness, so that goodness can then be defined in terms of function. Aristotle defines the natural purpose of a thing (*to pros ho pephuken*) as that for which the prudent man as such, and the appropriate knowledge, would use it (*Top.* VI 145^{a}25–7). This, however, is unhelpful (how would the prudent man use a tiger?), and may not even be intended to apply to anything but artefacts. In practice, Aristotle takes it that the widest or most general kind of thing which all non-defective members of a class can do, which differentiates them from other members of the next higher genus, is their function. This is the principle he uses to establish the functions of plants and animals in *De gen. an.* I 731^{a}25–^{b}5, and the function of man in *E.N.* I 1097^{b}33–1098^{a}4; and it does not involve the difficulties of Plato's definition: men can indeed detonate bombs, but the most general description of their behaviour which will not apply to the behaviour of other animals is 'acting from deliberate choice' or something like that, so that will be their function.

Clearly a function in this sense can be ascribed to things other than artefacts, and indeed one way, and perhaps an indispensable way, of defining any physical thing is to specify its function in this sense, its characteristic behaviour. That a thing's function in this sense is what is best for it, and what its parts and other features are for, might be argued as follows. When we say that a physical object is a good *f*, we do so on the ground that it has the features which are good for an *f*, and we call a size, shape, temperature, place, change, or the like good for an *f*, on the ground that having such a feature or undergoing such a change will enable it to do something, promote some behaviour in a wide sense on its part. Thus we might call a west wall good for a camellia on the ground that it will flower longest and most abundantly on a west wall. The word 'good' is used in connection with physical things always with an eye to or in relation to some behaviour. Now if we use the word 'good' in connection with an *f* relative to *ϕ*ing, we cannot consistently deny that

ϕing is good for *f*s: on the contrary, ϕing must be precisely what is good for *f*s, as that bodily state with an eye to which we use the word 'healthy' is precisely what is healthy. Sometimes, however, we may feel doubt whether ϕing really is good for *f*s, and if so, we will modify our claim that the features which promote ϕing are good for *f*s. A Greek of the fourth century might have said that sixteen hands is a good size for a horse because it is the best size for carrying a warrior in battle, and black is a good colour, because it is the best colour for striking terror into the hearts of enemies in battle, and so on: assessing features of horses relatively to carrying warriors in battle. If we asked him whether it is good for a horse to carry a warrior in battle, and this question shook him, he would have to say that it is only from a particular point of view that sixteen hands is a good size for a horse, and black a good colour. Now it will be found that these qualms arise in proportion as ϕing is more removed from the function of an *f* as defined above; where ϕing coincides with the function of an *f*, the features which enable an *f* to ϕ are said to be good for an *f* without qualification.

From this it seems to follow that parts of a thing, changes it undergoes, etc., are assessed as if they were for its function or end. An animal's eyes are good if they enable it to see the things it needs to see to perform its function; a plant grows well if it grows to be such as to live a long time for a plant of its kind and have a large progeny. Whether it is correct to say that 'the other things' *are* for the end, is equivalent to the question whether they are to be explained by the thing's form. If a thing has legs because it is a dog, i.e. the sort of thing to get its food by locomotion, its legs are for locomotion; and if it has legs for locomotion, it is not just flesh and bone disposed thus, but something which flesh and bone disposed thus constitute.

In **$195^{a}25$–6** Aristotle says that what a thing is for may be called good or apparently good. His point is that a jemmy may properly be called for housebreaking or a dog's struggles for escaping the veterinary surgeon, because these ends appear good respectively to the burglar and to the dog.

So much on the four senses in which, according to Aristotle, a thing can be called a cause or said to be responsible for something. As was said above, p. 55, things are called causes in any one of these senses, not because they have something in common, or because they conform to a single definite idea, but by analogy: we have no single idea answering to any of the of the expressions 'matter', 'form', 'source of change', 'end', and their meaning must be grasped in the way described above, pp. 72, 78.

Having distinguished the senses of 'cause', Aristotle goes on, in **$195^{a}27$–$^{b}21$**, to classify the logically different ways in which we can give the cause in any one of these senses. We may give the proper cause or something which is incidentally the cause (**$^{a}33$**); that is, we may give the cause under a description under which it is, or under a description under which it is not, responsible. Thus we might say 'Her necklace is made of gold and diamonds' or 'Her necklace is made of the spoils of the Incas'. It is because the spoils of the Incas are gold and diamonds, not because they once belonged to the Incas, that they are such as to be the material of the necklace. Or we might say 'He is in a hurry to marry the beautiful actress' or 'He is in a hurry to ruin himself'. His actions are for the objective marrying the actress under the description 'marrying the beautiful actress', not under the description 'ruin'. Again, the description under which we give a cause, whether appropriate or not, may be general or specific (b**13–15**). We may say 'The box is made of mahogany' or 'of wood', 'of the produce of Honduras' or 'of the produce of foreign climes'. Again, we may give appropriate and inappropriate descriptions either separately or together (b**10**). In the remark 'He was killed by a stab in the dark', whether the cause is given under an appropriate description or under a combination of appropriate and inappropriate, depends on whether it made a difference to the efficacy of the blow that it was struck in the dark. That of which any kind of cause is a cause may be given in the same variety of ways (b**6–7**). ('The marble was made into a statue', 'The marble was made into the worst eye-sore in Europe', 'He spent all his time in her company for love of her', 'He wasted the best years of his life for love of her'.) And, finally, a description of any of these kinds may be one under which a thing is a possible, or an actual, cause or thing caused (b**16**). As often (e.g. *Met. M* 1087a 15–18), Aristotle connects possibility with generality and actuality with particularity. Leather is a possible material for a shoe, but not for any particular shoe; it is the actual material only of particular shoes; similarly spherical is a possible shape for bronze; it is the actual shape, not of bronze, but only of some particular lump of bronze.

The classification of kinds of cause has received more attention than the classification of ways in which a cause may be given, and I shall say more about it when Aristotle has finished expanding it in chapter 7. The latter classification, however, is not without interest in its own right: the distinction between giving a cause under a description under which it is, and giving it under a description under which it is not, responsible ('He did it all for money', 'He did it all for nothing') seems akin to the distinction between intensional and non-intensional statements (see, for instance, J. O. Urmson, 'Criteria of intensionality', *Proceedings of the Aristotelian Society*, supplementary volume for 1968); and it is also essential to the argument of the following chapters. As Wieland observes (pp. 257ff.), Aristotle's argument for teleological explanation presupposes his analysis of chance, and in analysing chance Aristotle makes use of the distinction between proper causes and causes by virtue of concurrence.

CHAPTERS 4–6

These chapters constitute a fairly straightforward treatise on chance. In chapter 4 Aristotle raises the question whether there is any such thing, and considers some views of his predecessors; in chapter 5 he tries to say in general terms what chance is; and in chapter 6 he distinguishes two kinds of chance, luck and what he calls *to automaton*. I translate this 'the automatic', in preference to the more usual 'spontaneity', which has misleading connotations of acting out of free will. Luck turns out to be a subdivision of the automatic, and what Aristotle is trying to analyse under the name of the automatic is simply chance in general.

Chapter 4 presents no difficulties. Some people have wondered whether there is such a thing as chance; but all men, though accepting the venerable principle (**$196^{a}14$**: see textual note) that everything has a cause, nevertheless attribute some things to chance. Hence the earlier physicists ('they' of **$^{a}16$**) should have discussed it, particularly as they invoke it as an explanatory factor: so Empedocles (**$^{a}22$–3**) in his poem on the universe 'concerning nature' (DK 31 B 53), and Democritus (**$^{a}24$–$^{b}5$**), who made the universe, or rather the infinitely many universes, the result of a swirl among the atoms which was itself fortuitous (for a discussion of Democritus' view see C. Bailey, *Greek Atomists and Epicurus*, pp. 139–43). If, as Democritus held, the universe is the outcome of chance but plants and animals are not, this fact deserves comment (on **$^{a}36$–$^{b}1$** see textual note). That chance is a genuine cause but a kind of deity or supernatural being (**$^{b}5$–7**) is an opinion which may have been held rather by the ordinary man than by any notable thinker; though slightly irresponsible supernatural beings (*daimones*) come to play an increasing role in the government of the universe in later Greek thought (see the passages listed by Miss de Vogel, *Greek Philosophy* III, index, s.v. 'demons').

Aristotle begins his own account in chapter 5 by considering what sort of things are attributable to chance; for since all men do attribute things to chance, it is pointless to deny that there is any such thing (**$196^{b}13$–17**). Chance, says Aristotle, is not responsible for things which come to be always or for the most part in the same way (**$^{b}10$–11**). By things which come to be always in the same way he means, I think (see below, pp. 115–16), the natural behaviour of basic kinds of matter, such as the upward movement of fire, and things which are usual might be more complex phenomena like snow in winter (cf. $199^{a}1$–2). It is true that we do not ordinarily ascribe such phenomena to chance, and we should bear it in mind that when Aristotle denies that the behaviour of the heavenly bodies is due to chance, he need be saying no more than that it is like the natural behaviour of fire and earth (cf. *De caelo* I. 2).

The statement that things which do not come about always or usually in the same way are due to chance, and *vice versa* (b**15–17**), should probably be taken only as a first approximation to the truth, to be refined in what follows.

Aristotle next distinguishes a class of things which he calls 'for something', i.e. for the sake of something (b**17**). These are defined in b**21–2**, and the definition is important, as things which might have been done as the outcome of thought or nature. Things which are the outcome of thought would be rational actions, artefacts, and also things like health in a patient who has been properly treated and no doubt agricultural achievements like cattle-breeding, wine-growing. By nature is meant nature in the sense of form. Aristotle has not yet proved that anything is due to nature in the sense of form, but we may perhaps allow him to group together the apparently appetitive movements of animals and the growth of animals and plants including the formation of their organic parts as things which *might* be the outcome of nature. He would have exposed himself less to misunderstanding if he had called this whole class the class of things which *are such as to be* for something, rather than the class of things which *are* for something.

The logic of the remark b**19–21**, 'Clearly, then, also among ... belong to be for something', is questionable. If Aristotle thinks that from the premisses 'some things which come to be are unusual' and 'some things which come to be are for something' he can infer that some things which come to be are both unusual and for something, or even (an inference which Ross considers less unhealthy) that being for something is a possibility for some things which are unusual, he ought (cf. *An. pr.* I 29a6–10) to know better. Still, the conclusion is hardly disputable. The things which are most obviously for something, things due to choice, are 'things which come to be neither necessarily nor for the most part' (cf. *E.N.* III 1112a21–6). It is tempting to take the words 'things which come to be' in b17 to refer only to this class: chance has been found to lurk somewhere in it, and we would expect Aristotle rather to continue his search by subdividing it than to make a fresh start and try to prove that there are things which are both for something and unusual. But the words 'Clearly, *then, also*' (b19) are against this interpretation.

The things we ascribe to chance, Aristotle continues, b**23–4**, are such things of this sort (unusual and, in his extended sense, for something) as come about by virtue of concurrence, i.e. things which, though they might have been done for something, in fact were not. At first it may seem that this is too narrow, that a thing might be ascribed to chance even if it is not such as to be for something; but on reflection we may decide that Aristotle's instinct is sound. We ascribe a thing to chance only if we think it remarkable, and it is doubtful whether we should think a thing remarkable, doubtful whether we would even notice it or be able to pick it out from the rest of our environment, if it did not seem to us, at least in a weak sense, such as to be for something. There are many shapes it is as improbable a pebble should have as that of a sphere; but we should notice a perfectly spherical pebble on the beach, because it would do as a marble, because it satisfies our taste for symmetry, because we can expound its geometrical properties, and so on. To take a more fundamental case, an infant differentiates an object from the rest of the world because it can move the object, wave it, drop it, etc. The object is such as to be manipulated and played with, such as to produce a bang or thump. This is an extension of the point to which we shall return below, p. 124; for the moment it seems correct to say that nothing is attributed to chance (and still less to luck) unless it is in a fairly striking way such as to be for something, adapted to some end.

Aristotle is now in a position to define chance as a cause by virtue of concurrence of things which come to be neither always nor for the most part, and are such as to be for something (**197a5–6, 32–5**). This definition is, I think, largely sound, but the crucial passage explaining it, **196b24–197a5**, needs some clarification.

First, what exactly is it in the example in **196b33–197a5** which is due to chance? *A* is owed money by *B*, and going to a place to which he does not normally go, finds *B* in funds (perhaps because *B* is collecting contributions for a club dinner: Aristotle seems to be referring to a story well known in his day but now lost) and thus recovers his debt. There are two things here which might be called lucky, recovering the debt, and going to the place where *B* is. Most commentators have taken it that it is the recovery of the debt which Aristotle regards as the outcome of luck, but this does not fit Aristotle's account of a thing due to luck: it is not something which might have been done for something but in fact was not. I prefer, then, to say that it is *A*'s going to where *B* is, which is the outcome of luck. This does fit the general account of lucky outcomes, as is shown in 196b33–4: if *A* had known that *B* would be there, he would have gone there for the sake of recovering the loan. Further, in 196a3, where Aristotle is apparently thinking of the same story, it is the going to the market-place which is ascribed to luck. Similarly in 199b20–2, the family friend arrives on the scene by luck: he acted as if he had come for paying the ransom, but that is not why he in fact came. Similarly, too, *Met.* Δ 1025a25–30. And again, on the traditional interpretation (see Ross on 196b36–197a1) it is not strictly relevant that *A* did not go to the place normally; but Aristotle is an economical writer, and if it is the going to where *B* is which is lucky, there is some need for this addition. It must be admitted, however, that Aristotle's treatment of the example is not very careful. Even if *A* regularly goes to the market-place, he may still be lucky to come across *B*, so long as *B* does not regularly go to the market-place, and their paths do not normally intersect elsewhere. It is *A*'s going to where *B* is which must be unusual. Further, Aristotle takes it that *A* is not actually looking for *B*, not wandering about for the sake of collecting the debt. Even if *A* does go to the market to find *B*, his finding him might be a matter of luck, if *B* seldom goes to the market. What would be due to luck here would be less *A*'s going to where *B* is, than *B*'s going to where *A* is, and this would be rather *B*'s bad luck than *A*'s good.

Aristotle is a little careless also in his talk about concurrence: he says that chance is a cause in the way in which a pale flute-player may be the cause of a house (**$196^{b}26$–7, $197^{a}14$–15**). There is, in fact, an important difference between a thing due to chance and a house built by a musician. A house, under that description, i.e. *qua* house, has a proper cause, a builder (for the musician knows how to build). A thing due to chance, as such, i.e. under the description under which it is attributed to chance, has no proper cause, though under another it may have and Aristotle thinks ($197^{a}10$–11) it must have. Thus if *A* is not looking for *B*, nothing is the cause of his going to where *B* is; and nothing is the cause of the tile's falling on the pedestrian ($197^{b}30$–2); but there is a cause of *A*'s going to the market-place, and of the tile's falling in the path it does. It is precisely because some things which are adapted to an end, as such have no proper cause, that they are ascribed to chance. It would be more correct, then, to say that a thing due to chance is a concurrent outcome, than to say that chance is a concurrent cause.

In **$197^{a}8$–32**, Aristotle shows that his account confirms and clarifies the respectable opinions or *endoxa* about chance collected in chapter 4 (for the importance of doing this see above, pp. ix, xv). We can see why chance is indefinite and inscrutable (**$^{a}9$–10**), incalculable (**$^{a}18$**), and inconstant (**$^{a}30$–2**). Aristotle's account accords with and illuminates ordinary talk about good and bad luck (**$^{a}25$–30**). More important, we see that there is no conflict between the theses that everything has a definite cause, and that some things are due to chance (**$^{a}10$–14**), for (this point might have been made more sharply) the same thing under one description may have a definite proper cause, and under another be due to chance. Thus under the description 'going to the market-place' what comes about has the proper cause 'wishing to go marketing' ($196^{a}4$–5); under the description 'going to where his debtor is' one and the same occurrence has no proper cause, and is attributed to chance. Chance, then, is not something over and above proper causes, but is always some proper cause or other, in relation to an outcome by virtue of concurrence.

$197^{a}21$–5 are difficult lines, partly because the text is uncertain, but partly because they seem irrelevant. In *Met. Z* $1034^{a}10$ we are told that a man can become healthy by luck. The idea is that equalization of temperature in his body (cf. $1032^{b}18$–19) might have been effected for health, but is in fact the incidental outcome of something else. It does not seem relevant to the discussion of chance (though it might be to the discussion of causes generally) whether we say that it is fresh air or a hair-cut which plays the part of chance here. If we adopt a couple of variant readings, *εἵλησις* for *εἴλησις*, and *ἀποκεκαθάρθαι* for *ἀποκεκάρθαι*, Aristotle might be saying that some things ought not to be ascribed to chance because there is some factor which can plausibly be held responsible as a proper cause, and confusing this point by combining it with an expression of scepticism about mystical health cults: 'Could breathing or the revolution of the heavens be the cause of health, and not having had a laxative?'—but this is probably too far-fetched.

In chapter 6 Aristotle goes on to distinguish chance into luck and the automatic. Only rational beings can be lucky, and hence it is only when that which is due to chance is done by a rational being that it can be ascribed to luck. More precisely, that which is the outcome of luck must be something the doing of which might have been rational activity (**$197^{b}1$–6**). (The word translated 'rational activity', *praxis*, is important in Aristotle's moral philosophy, and has been the topic of much recent discussion; Aristotle seems to use it for an action or activity in so far as that action or activity is the exercise or realization of some rational disposition.) Hence an inanimate object cannot do anything lucky, but might have something lucky happen to it (**$^{b}11$–13**), for instance if a sculptor was distracted and his chisel slipped and made a groove which gave the face of the statue just the right expression.

Aristotle allows the possibility of good and bad luck, but does not tell us what would be ascribed to the latter. Perhaps the doing of something the not doing of which might have been rational activity. Thus suppose the debtor of $196^{b}33$ ff. is a dishonest character; he might call it bad luck that he went to where *A* was, since if he had known that *A* would be there, he would have kept away.

If that which is lucky is something which—or the opposite of which—might have been due to thought, we might expect the automatic to be something which (or the opposite of which) might have been due to nature. Aristotle's examples hardly answer this expectation. In **$197^{b}15$–16**, a horse loses its rider in battle, and wandering about meets its groom. It is possible that going to its groom would have been a natural thing for it to do; but we may think that this is more a case of something which a man (the rider or groom) might have brought about deliberately. And the other two examples are certainly of things which might have been due rather to thought than nature. A three-legged stool is thrown up in the air and lands on its feet in such a way that a man could sit on it (**$^{b}16$–18**); the tile which falls off the roof and hits someone does so as it would have done if it had been thrown for that purpose (**$^{b}30$–2**).

These examples suggest that a chance doing or undergoing is always one which might have been due to thought, and it is lucky if done or undergone by a man, and 'automatic' if done or undergone by anything else. The summary of the distinction (**b18–22**) suggests a similar interpretation. Aristotle seems there to be connecting the distinction between lucky and automatic occurrences with the distinction between things which would have to be brought about by an external agent, and things which could be chosen. Though a man can choose to lead a horse to safety, he cannot choose that it will find its way to safety, in the way he can choose to go where his debtor is. (Or, if we think that this contrast is a little forced, and also prevents luck from being a subdivision of the automatic (**197a36–b1**), we could take the phrase 'has an external cause' to mean 'is not in fact due to nature or mind'; luck will then be distinguished by being 'for something capable of choosing'.)

Although, however, Aristotle seems to overlook the case of that which might have been due to nature in the sense of form, but not to thought, this case becomes of great importance in chapter 8, and it is hard to think he would wish to deny it is a genuine case of the automatic. Perhaps it is not considered here because it has not yet been proved that anything is ever due to nature in the sense of form but not to thought.

Aristotle concludes the discussion of chance with some riders which are difficult. The Greek for 'it in vain' is *auto matēn*, which looks like *automaton*, automatic, but whilst Aristotle makes play with this verbal resemblance in **197b29–30**, it is not clear what point it is supposed to illuminate. He might mean either (i) when something is in vain, that is the reverse of the automatic. It is an automatic outcome when something not such as to produce an end produces that end; something is in vain, when it *is* such as to produce an end, but fails to produce it. Or (ii) in a case of the automatic, something not such as to produce a good end produces something good; when something not such as to produce a bad end produces something bad (the automatic working *despite us, matēn genētai*), that is like when something is in vain. The second interpretation is more complicated, but perhaps fits Aristotle's words better.

No less perplexing are lines **197b32–7**. Ross thinks the reference is to spontaneous generation (for Aristotle's account of which see *De gen. an.* III, 11), and interprets Aristotle as saying that this shows us most clearly what the automatic as contrasted with the lucky is like, though it is not an automatic outcome of the type described above. Aristotle would certainly do well to mention this further class of automatic occurrences, but it is hard to get Ross's sense out of the text. I am more inclined to interpret: 'That which is in accordance with nature (in the sense of form) is (contrary to the opinion of people like Empedocles, see below, 198b16–32) the very last thing we should attribute to luck. For when something is contrary to nature (e.g. a deformity like webbed fingers) it is not called the outcome of (bad) luck, but rather an automatic outcome, and even that is not correct. An automatic outcome has an external source, and here the source is internal (e.g. defective seed, cf. 199b6–7).'

Finally, there is the argument of **198a5–13**. Even if the heavens are due to chance, they are also, and more strictly, due to nature or mind. For whatever is due to chance is also, under another description, due to a proper cause, and this proper cause will be nature or mind. So a3–4 and, perhaps, see the textual note to a6–7, a7. Aristotle's expression here is awkward, because the nature to which a chance outcome might have been due is nature as form, and the nature to which it is in fact due may be nature as matter.

CHAPTER 7

In chapters 4–6 Aristotle has argued that chance is not a further kind of cause over and above the four distinguished in chapter 3; in this chapter he restates and rounds off the fourfold classification.

The chapter begins with a summary argument to show that there are just four types of cause: there are four ways of asking and answering the question 'on account of what?' (**198a14–21**). This argument seems to indicate a fresh approach. Although the question 'on account of what?' is mentioned in chapter 3 (194b19, 33), the causes are there presented chiefly as answers to the questions 'What is it made of?' 'What is it?' 'What was the source of change to it?' 'What is it for?'—questions different in form, which may all arise over *things*. The question 'on account of what?' never arises over things in the same way: it arises over happenings or facts. We do not ask 'On account of what is a statue?' but at best 'On account of what is this a statue?' or 'On account of what do statues exist?' Hence in so far as the causes are given in answer to questions of the form 'On account of what?' they will be responsible less for things than for facts or happenings.

Now can causes of all four types be responsible (though in different ways) for the same happening or fact? Do they supply answers to the questions 'On account of what, as matter, is it the case that *p*?' 'On account of what, as form, is it the case that *p*?' etc.? In *Met. H* 1044a34–b1 Aristotle envisages questions like these arising over things: we should ask, he says, what is responsible for a man as matter, what as form, etc. But it is not easy to see how something could be responsible as matter for a fact or happening, and Aristotle's language in 198a14–21 suggests that he thinks that the question 'On account of what?' is a demand for different kinds of cause, not in so far as it is tacitly qualified 'On account of what as matter' etc., but according as it arises over different kinds of fact or happening. Some facts or happenings can be explained only by a formal factor, others only by a material factor, and so on.

This is an interesting suggestion, especially because even if factors of all four types are responsible in a general way for a single thing, say a man or house, we would expect each to be responsible for a different aspect of or element in the man or house. However, the suggestion is not worked out, here at least, with any rigour. We are told that the facts of mathematics (e.g. the fact that the diagonal of a square is incommensurable with the side) are to be explained by formal causes, that is, by what a straight line is, by what commensurability is, etc. (^{a}**17–18**). Mathematics, however, is different from the study of nature, and we are left in the dark whether anything which is the concern of the student of nature needs to be explained by the formal cause. Then we are told that a war (which is presumably a process or series of happenings) can be explained either by a source of change or by an end (^{a}**19–20**), which suggests that there is no logical difference between things explicable by the one sort of factor and things explicable by the other. Finally, Aristotle most cryptically says that when the question 'On account of what?' arises over 'things which come to be' (better than simply 'happenings', though the text might be taken either way), it is a request for the matter (^{a}**20–1**). If a living thing or an artefact is a thing which comes to be, surely Aristotle's view is that it is to be explained by causes of all four kinds (cf. ^{b}5–9).

Having re-emphasized, at any rate, that there are these four types of cause, Aristotle goes on to say (^{a}**21**–^{b}**9**) that the student of nature should familiarize himself with, and be prepared to give, all four. Some of the details of this passage, as well as its general trend, present difficulty.

In ^{a}**24–7** Aristotle says that three of the causes (i.e. form, source of change, and end) often coincide. The usual interpretation of this is that the efficient cause is a form operating *a tergo*, and the final cause a form operating *a fronte* (so, for instance, G. R. G. Mure, *Aristotle*, ch. 1). This interpretation fits Aristotle's statement that 'a man gives birth to a man' (^{a}26–7), but otherwise seems unsatisfactory. Aristotelian causes do not so much operate, exert what Hume calls power and efficacy, as provide explanations, and tend to be not before or after, but contemporaneous with, the things for which they account (195^{b}17–18, see above, p. 101). In *De an.* II 415^{b}8–27 Aristotle says that the soul stands to the body as cause in these three ways, and the *a tergo–a fronte* metaphor does not apply well to the soul. And, as we shall see more fully, in many cases form, source of change, and end coincide because when a form is a source of change, it is a source of change *as an end*.

In ^{a}**28–9** Aristotle refers to unchangeable changers. In ^{b}**2–3** he identifies these as 'that which is completely unchangeable and the first thing of all, and a thing's form'. That the former, God, is no concern of the student of nature (^{a}28), is perhaps clear, but it is surprising to be told that the form is not natural (^{a}**36**), but falls outside the field of natural science because it has no change or source of change in itself (^{a}**28–9**, ^{b}**1**): the burden of chapters 1–2 seemed to be that a thing's form *is* a source of change and *is* the concern of the student of nature. Perhaps Aristotle is recurring to the point made in 194^{b}9–15, that the student of nature should not concern himself with forms in separation (see above, pp. 97–8), and is saying that what appears separately in account, though a source of natural change (198^{a}36), has no source of change in it (as have 'natural forms which pass away', like a man, a dog), but is responsible for change as an end (^{b}3–4).

198^{b}5–8 are also obscure. Ross thinks that an explanation of the type 'this out of this necessarily' is an explanation by the source of change, and an explanation of the type 'if so and so is to be (as the conclusion out of the premisses)' is an explanation by the material cause. His reason is that the premisses are said to stand to the conclusion as material cause (195^{a}18–19). That is true, but the parenthesis '(as the conclusion . . .)' does not illuminate the phrase 'if so and so is to be', and I should like to see it moved back a line, so that the passage reads: 'this out of this necessarily, as the conclusion out of the premisses (but it may be out of this simply, or out of this for the most part); and if so and so is to be; and this would be what the being would be', etc. 'This out of this necessarily' is a natural expression for the material cause (cf. 195^{a}19), but a perhaps unparalleled one for the source of change (the use in *Met. Δ* 1023^{a}30–1—a fight arising out of insults—is not a very impressive parallel). 'If so and so is to be' is not a usual expression for the source of change either, but the things which must be done or occur if some end is to be achieved might be said to explain the attainment of the end as source of change. Thus what a carpenter does to a piece of wood explains the coming into being of something which can do the work of a rudder (cf. 194^{b}5–7) as source of change. At the same time we should notice that Aristotle often says that the matter, e.g. bricks in the case of a house (200^{a}24–6), is that which must be 'if so and so is to be'. It is a feature of Aristotle's doctrine of the causes, which we shall consider more fully below, that he connects the source of change and the matter very closely together.

Aristotle so emphasizes in 198^{a}21–^{b}9 the importance of knowing about and seeking all four kinds of cause, that he may seem to think that every natural phenomenon has causes of all four kinds. That this is not his opinion appears from *Met. H* 1044^{b}8–12: 'Things which are due to nature, but which are not realities, do not have matter, but the underlying thing is the reality. Thus what is the cause of an eclipse—what is the matter? There isn't any, but the moon is the thing affected. What is responsible as source of change and destroyer of light? The earth. What is it for? Perhaps it is not for anything.' Aristotle's position, then, is that the student of nature should always look for explanatory factors of all four types, but he cannot always be sure of finding them; they will all be found, perhaps, only in connection with living things.

It is now time to consider what we are to make of the doctrine of the four causes as a whole. As Aristotle himself presents it (herein unlike some of his commentators), it is extremely flexible. On the one hand, logically quite heterogeneous entities are grouped together because their explanatory role is rather of one type than of another. Thus strength is grouped with a sculptor (cf. 195a9, 34) because it is responsible for manual labour rather as the sculptor is responsible for the statue than as the bronze is. On the other, the types of cause are not mutually exclusive: the same thing can be a cause in more than one way. In chapters 1–2 Aristotle considers the claims of form and matter to be called nature. If by nature he means a source of change, he is writing, not as if the source of change must be a factor over and above the form and the matter, but as if it were reasonable to ask which of those two factors it is. Similarly he writes elsewhere, not as if the form were a factor over and above the source of change and the end, but as if it were reasonable to ask whether the form of a particular thing is a source of change or an end. Thus in *Met. Z* 1041a27–30 (retaining with Ross the MSS. reading), he says that 'what the being of a thing would be' 'is in some cases what it is for, as, perhaps, in the case of a house or bed, and in others the primary' i.e. proximate 'changer'. The text here has been doubted, but the point is sound—many expressions like 'murder', 'punishment', 'ploughing', and (possibly) 'tide' are used not just for physical changes, but for physical changes effected for something and/or by something—and Aristotle's practice accords with it. To say that a house is a shelter preventive of destruction by wind, rain, and sun is to give the form of a house (*De an.* I 403b1–5, cf. *Met. H* 1043a16–18), and also to say what it is for; to say that a lunar eclipse is a screening by the earth is to say what it is (*An. po.* II 93b5–7 etc.), and also to give the source of change. Aristotle's view seems to be that where a fairly complex thing is for something, to know what it is is to know what it is for, and where it is not for something, to know what it is is to know the source of change. Again, Aristotle often seems not to discriminate between the material cause and the source of change. The carpenter of 194b4–7, who knows by what changes or carpentering operations a rudder is to be made, surely does know the source of change, but is said to know the matter; in 200a31–2, the matter and its changes (which surely stand to the end-product, whether of nature or of art, as source of change) are grouped together. Similarly with the formal cause and the end: we are often told that they are one and the same (e.g. *De gen. an.* I 715a4–9), and when they coincide with the source of change as well, that is often, as suggested above, p. 112, because the form is a source of change as an end.

Although, however, the fourfold classification of causes is flexible in these ways, underneath it there seems to lie a firmer twofold division between two radically different types of explanation, one employing the concepts of matter, source of change, and unconditional necessity, the other employing the concepts of form, end, and conditional necessity.

Sometimes a phenomenon can be explained as the direct outcome of unconditional necessity, and in that case it is ascribable simply to matter. In *An. po.* II 94b37–95a3 Aristotle says: 'Necessity is of two sorts: that which is in accordance with nature and tendency (*hormē*), and that which is violent and contrary to tendency. Thus a stone necessarily moves up and down, but the necessity is different in the two cases.' The necessity with which it moves down is what we may call unconditional necessity, and what is unconditionally necessary can be explained simply by specifying the matter (cf. 200a1–5); the stone goes down because it is made of earthy stuff. Most of the phenomena in which we are interested, however, are more complex. If the moon is made of stuff which naturally moves in a circle (for the existence of such stuff, see *De caelo* I. 2), it will be the direct outcome of necessity that it goes round the earth; but since it is not natural (but rather an automatic outcome) that it passes through the earth's shadow, eclipses are not a direct outcome of necessity, and have to be explained by an external source of change, the earth. However, and this is an important point, an external source of change by its action accounts for the change in accordance with the nature in the sense of matter of the thing affected. That is most clearly seen over artefacts (though they, of course, can also be explained in a different way altogether): the maker of a saw by acting on iron is responsible for the coming into being of a saw, but because that on which he is acting is iron and not wood or wool (*Met. H* 1044a27–9). But the same applies to natural things which are not for anything: it is because fire is as it is that when extinguished in clouds it makes the noise called thunder (cf. *An. po.* II 94b32–3, *Meteor.* II 369a24–33). Confirmation of this can be derived (if its Aristotelian authorship or authenticity is allowed) from *Meteor.* IV: 'natural' change (cf. 198a28–9) is defined as change due to or explicable by the character of a thing's material constituents (378b31–4), and a host of phenomena like boiling (380b13–14), the floating of wood (384b15–18), are explained directly by what the things affected are made of. A firm distinction is drawn between things to which this sort of explanation is applicable, viz. homoeomerous things like flesh and silver, and things to which it is not, e.g. heads and loving-cups (390b2–14). Compare also *De gen. an.* II 734b31–735a2, 735b16–21, 26–37, V 778a29–b1, b10–19, 789b2–15. In general, changes which Aristotle attributes to an external source are changes explained in accordance with the laws of physics and mechanics, and, since these are laws governing or describing in general terms the behaviour of the elements, this is as much as to say that they are changes due to nature in the sense of matter.

In the light of this we may perhaps better understand some of the remarks noted as difficult above: that the carpenter's knowledge of how to make the rudder is knowledge of the matter (194b4–7), and that the matter is the cause in the case of things which come to be (198a20–1). (With that, compare the remark, *Met. Z* 1034a11, that the matter can be 'in control of' (or 'the beginning of') the coming to be of things which can also come to be by nature or art.) Again, Aristotle's expression for the material cause in *An. po.* II 94a21–2, things such that 'if they are, this necessarily is' (see above, p. 100), may be inspired by the idea that the matter and its changes render the thing to be explained unconditionally necessary.

This conception of explanation may remind us of modern regularity theories. The kind of necessity involved is the necessity attaching to 'things which are thus and by nature such as to be thus' (*De part. an.* I 642a34–5), or to 'things which are always the same' (cf. *Met. Δ* 1026b27–8). There is no reason why a thing which is thus of necessity in this sense, should be thus: if there were, the necessity would no longer be unconditional. It is a feature of this kind of explanation that there should be no reason for the conjunction of the action of the source of change with the phenomenon to be explained. And Aristotle does not seem to insist that the law in accordance with which the source of change explains the phenomenon must hold in all cases: it is enough if it holds for the most part. He speaks in the same breath of that which is of necessity, i.e. always the same, and that which is for the most part, e.g. 198b5–6 *Met. Δ* 1025a15, *E* 1026b35, as if they play the same role in scientific explanation (1027a20–1, cf. *An. po.* II 96a8–19). Although, however, explanations of this type are guaranteed only by natural laws to the effect that things always or usually happen thus, to give such a law is not, for Aristotle (or, perhaps, for Plato: cf. *Phaedrus* 270 c–d), to state a merely contingent observed regularity: it is to explain the nature, in the sense of matter, of the thing affected. If the kettle did not boil when the flame was under it, we would not say that the water in it was behaving irregularly: we would say that what was in it was not water (for water would necessarily have turned to air in such circumstances); and if no law could be found in accordance with which the contents of the kettle behaved, the contents of the kettle would be completely unknowable.

This is the way to explain anything which is not for anything, and we should try to explain even things which are for something, like breathing (*De part. an.* I 642a31–2) and artefacts (otherwise we should not be able to make them), in this way too. But where a thing is for something, it can also be explained by what it is for. Thus we can say: tigers must have teeth the size and shape they have; otherwise they would not be able to bite (cf. *De part. an.* III 661b25); an axe must be the size and shape it is; otherwise it will not cut down trees (cf. *De part. an.* I 642a10). In these cases the thing explained is the outcome, not of unconditional, but of conditional necessity, and it is important to gauge the force of this correctly. On the one hand, it is conditional: it is necessary that tigers should grow teeth like this, not absolutely, but *if* they are to bite well. On the other hand, it will not do to explain just anything as being conditionally necessary. We should not, for instance, explain the sharpness of just any old piece of flint we pick up, as necessary if it is to cut. The sharpness of flints in general would be a good example of something to be explained by the matter: they are sharp because formed in this way out of this kind of stuff. (Nor, if we say with Kant that objects of experience must be spatial and stand in causal relations, since otherwise experience would be impossible, should we think we have *explained* the presence of these features in objects of experience, unless we also say with Kant that objects of experience are formed by the mind for the sake of being experienced.) We can use the end 'biting' to explain the size and shape of these teeth only (if at all) if they grow as parts of a tiger, and we can use the end 'health' to explain a post-prandial walk, only if taking it is following a doctor's advice or a precept of medicine. It is an idea fundamental to chapters 4–6 and 8, that a thing can quite easily be such that it might have come to be for some end, but whether it can be *explained by* that end depends on whether it comes to be as an outcome of art or nature in the sense of form, or as an outcome of chance and unconditional necessity. Aristotle's position seems to be that a thing can be explained by an end only if its coming to be is an exercise of some sort of knowledge or skill, or the realization of some disposition—for nature (*Met.* Λ 1070a12) is a sort of disposition.

The distinction between these two modes of explanation resembles the distinction sometimes drawn today between explanation by causes and explanation by reasons. A cause is an event or circumstance which explains what is to be explained in accordance with physical laws; physical laws being general statements of what always or usually happens. Human actions, it is thought, are not, and cannot be, explained in this way. They must be explained by circumstances which constitute reasons for them, which render them intelligible in the sense that we can see them as reasonable, or at least see the point or good of them. Now physical laws hold for or relate to quantities of matter; so if a thing's behaviour is explained by a cause, it may be said to be due to its matter. On the other hand, if something is explained by a reason, so that we can see the point or good of it, this is the sort of thing which Aristotle says is due to a thing's form. We might put the matter thus: both types of explanation appear to be of the pattern: factor C explains the behaviour ϕ on the part of f, f being such as to ϕ given C; but the explanation may be of two sorts. If it is in accordance with physical laws, then *C* is a cause, and *f* is what the thing ϕing is made of. If the explanation is one showing the point of the ϕing, or showing the ϕing as some kind of pursuit or avoidance, then *C* is a reason, and *f* is not some kind of stuff, but what an expanse of stuff constitutes.

Aristotle regards these types of explanation as partly but not wholly complementary. The behaviour of iron in taking on the shape and structure of a sword (cf. *De gen. an.* II $734^{b}37$–$735^{a}3$, I $730^{b}15$–19) is explained by action on it in accordance with physical laws; but the subjection of the iron to these forces is itself to be explained in a different way by factors relating to the form of the smith: the smith needs a sword (cf. *De motu* $701^{a}17$–22) and, as a smith, knows how to make one.

The modern reader may think that Aristotle is right to explain the coming into being of artefacts in this way, but wrong to attempt the same sort of explanation of the coming into being of living things and their organic parts. We shall see in the next chapter how he defends his procedure, but for the moment it is clear at least what his procedure will be. The chemical changes in what a tiger eats can be explained by the action of its teeth, stomach, etc., in accordance with physical laws (cf. *De gen. an.* II $743^{a}4$–8), but the subjection of food to these changes is to be explained by reference to other factors, which make the process intelligible in the sense that we can see the good of it (cf. *De part. an.* II $652^{b}13$–15). And as, when we explain the coming to be of the sword in this latter way, we see it as a realization or exercise of a disposition possessed by the smith, so when we explain the growth of a living thing in this way we see it as the realization of a sort of disposition possessed by the living thing itself. The difference between the two cases, Aristotle maintains, is that while the smith is primarily a man, while the concept of a man, as we might put it, is the concept under which we identify and re-identify a smith, and smithcraft is a disposition of a man, a tiger is primarily a thing with a disposition of which growth is a realization; there is no more fundamental description under which a tiger can be identified than 'thing which nourishes itself, pursues and avoids, etc.', nothing which this disposition is a disposition of; and hence rather than a disposition it should be called a nature or reality in the sense of form.

So much on the doctrine of the four causes as a theory of explanation; there is a slightly different way of using it suggested in *An. po.* II. 11, and it will complete our survey to look briefly at this.

In *An. po.* II. 11, Aristotle says that any one of the causes may function as a middle term or intermediate in a proof. This sounds promising. If propositions of different sorts are proved through different intermediates, then since that through which something is provable can be called responsible for it, we shall be able to pair off different kinds of cause with different kinds of fact or happening. Unfortunately the chapter belies its promise: Aristotle's illustrations are useless. His example of a material cause turns out to be a formal cause; his source of change is an event preceding the thing to be explained, and according to himself (see above, p. 101) a cause ought to be contemporaneous with what it explains; and his final cause fails to appear as a middle term at all.

Nevertheless, he could have made his point very neatly. He sets out his syllogisms, not in the 'traditional Aristotelian' way,

$$\begin{array}{c} M-P \\ S-M \\ \hline S-P, \end{array}$$

but with three terms in a straightforward sequence, which he calls the first, middle, and last, and for which he uses consecutive letters of the alphabet (for an expansive discussion of this see L. E. Rose, *Aristotle's Syllogistic*). In this layout we could see the first term as occupying the place for final causes, the third for material, and the second for sources of change or formal causes, according as it is subject or predicate. In first-figure syllogisms, then, where the first term is related to the second and the second to the third (*An. pr.* I $25^{b}32$–4), the source of change and the formal cause are exhibited as intermediates; e.g. loss of light belongs to screened by heavenly body, and screened by heavenly body to the moon (for the example see *An. po.* II $93^{a}30$–1 and ff.). The screening appears first as source of change—screening is the source of loss of light—and second as an affection of the moon. In the second figure, where the first term in the layout is the intermediate ($27^{a}5$–6), the final cause functions as an intermediate. For instance: why not sit around after meals? Health belongs to man as an objective, and does not belong to those who sit about after meals. Or again, why not pocket the spoons? Justice belongs to man as an objective, but does not belong to those who pocket other people's spoons. It might be thought that the same arguments could be presented in the first figure: not sitting around belongs to the healthy, and health, we wish, to us. But this arrangement would probably not please Aristotle so much. It is not the case that the healthy do not sit about after meals because they are healthy: if anything, the healthy can afford to linger over their coffee longer than the unhealthy; rather the situation is that health does not belong to those who sit around because they sit around. So we should argue in the second figure, with health as the predicate. Similarly it is actions which make us just or unjust, not the other way round (*E.N.* II. 1), so we should argue from 'justice does not belong to those who steal', not 'stealing does not belong to those who are just'. In the third figure, where the third term in the layout is the intermediate ($28^{a}11$–15), the material cause functions as intermediate. This figure is the one for proving that *A* belongs to *B* by virtue of concurrence. Thus, suppose we want to show that some musicians are pale—for the example and the point that the cause must be the matter see *Met. E* 1027a11–15—there is nothing in pallor to suggest knowledge of music or vice versa, so we must find something to which they both belong, and our argument might run: pallor belongs to all Yankees, and knowledge of music to some. Yankee stands as underlying thing to pallor and knowledge of music, and explains their concurrence.

On this showing, different causes function as middle terms in different syllogistic figures, and different syllogistic figures are suitable for proving different sorts of proposition. The first figure is the one for proving propositions in mathematics (*An. po.* I 79a18–19), and if, as I suggest, the formal cause is intermediate in the first figure, we have here the reason why formal factors are responsible for mathematical facts (198a16–17). The second figure is suitable for proving that some action or inaction ought not to belong (moral prohibitions), and here the final cause functions as intermediate. The third is suitable for showing that something belongs to something by virtue of concurrence, and in it the material cause functions as intermediate. Aristotle does not develop the doctrine of the four causes in this way, but it is one way in which it might be developed.

CHAPTER 8

Aristotle says in **198b10–12** that he will first give reasons for holding that nature is a cause for something, and then discuss the role of necessity in natural things. The former task is tackled in chapter 8, and the latter in chapter 9.

Chapter 8 is one of the most controversial in Aristotle. The general verdict since the Renaissance has been that Aristotle's use of final causes to explain natural processes is a disastrous mistake. Wieland, on the other hand, argues that Aristotle's teleology is completely innocuous, since it is 'als ob' in character and the notion of an end is a mere 'concept of reflection' to which nothing need correspond *in rerum natura* (pp. 261, 271, etc.). The general view seems to rest on a misunderstanding of the thesis that nature is a cause for something. Aristotle nowhere maintains that everything which is due to nature is for an end; on the contrary, as we have seen, he holds that things which are due to nature in the sense of matter are not for anything, but are just necessary unconditionally. What he maintains is that *some of the things* which are due to nature are for something. This is the sense of the cautious remark that 'the "for something" is present in things which are and come to be due to nature' (**199a7–8**, cf. 30, b10, 200a8). The things due to nature which Aristotle holds are for something are in fact the organic parts and the natural or unconstrained changes in respect of size, shape, place, etc., of plants and animals, and not even all of these: eyes, for instance, are for something, but they may not be blue for anything (*De gen. an.* V 778a30–b1, b16–19). (The whole plant or animal also comes to be and is due to nature, but this, as the 'natural form', is not so much for something as 'what the other things are for'.)

Whilst, however, it is a mistake to suppose that Aristotle's account of nature is teleological throughout, it is also wrong to suppose that where Aristotle thinks teleological explanation appropriate, he is not committed to holding that there is a basis for it *in re*. The thesis that some changes undergone by plants and animals are for something, is, as we have seen, and as Aristotle himself says in **199a30–2** (for a discussion of which see textual note ad loc.), equivalent to the thesis that they are due to nature in the sense of form, and the form of a thing is for Aristotle very much of a reality—is, indeed, what has the best claim to the title of 'reality'. If we ourselves shrink from saying that dispositions like a craftsman's skill are mere concepts of reflection to which nothing corresponds in the craftsman, Aristotle would resist even more strongly a similar suggestion about nature as form.

In assessing Aristotle's teleological thesis, we may do well to consider the background to it. It appears from Plato's *Laws* (X 889) that the current orthodoxy was something like this. Fire, water, earth, and air are natural, and move by necessity with their own characteristic movements; and all natural things like metals, stones, plants, and animals are due to their chance encounters and combinations. Such is the realm of nature; contrasted with it is the realm of art and mind, which is of comparatively recent emergence and very limited extent. Plato himself finds this picture inadequate, but is unable to liberate himself completely from the nature–art antithesis. In the *Timaeus* he represents some things as due to mind, others to necessity. It is notoriously unclear whether he thinks the works of mind were in fact effected by a personal demiurge, or (as Aristotle tends to interpret him) in some way by 'separate' forms and numbers themselves; but in general, the alternatives as Plato sees them are: things are due either to necessity, chance, and the natural movements of elements, or to mind, thought, reason. Perhaps many today would agree that this exhausts the possibilities; Aristotle, however, is trying to advocate a third: that living things have a source of change internal to them, which is distinct on the one hand from the nature of their matter and on the other from mind and skill. He replaces the antithesis of matter and mind with the antithesis of matter and form.

He begins with a lively sketch of the orthodox view, as held by Empedocles (**198b16–32**). This, though Aristotle gives it short shrift (b34), is often regarded as a brilliant anticipation of Darwin's theory of the evolution of species by natural selection (so Ross, *Aristotle*[1], p. 78).

In fact, it is doubtful whether Empedocles was any nearer to Darwin than Aristotle was. Empedocles did not think that life began with simple organisms, which were nevertheless as well adapted for survival as their simplicity allowed: he thought that there were first separate organic parts like neckless heads, totally incapable of survival (DK 31 A 72, B 57). Aristotle, on the other hand, is prepared to consider the possibility that men and quadrupeds first originated in the spontaneous generation of simple organisms (scoleces: *De gen. an.* III 762b28–763a5). Empedocles' suggestion that the division of the backbone into vertebrae is the result of random breakage through excessive bending in the womb (*De part. an.* I 640a21–2) would seem as absurd to Darwin as it did to Aristotle. And Darwin makes free use of the Aristotelian notion of conditional necessity. Many remarks could be taken from *The Origin of Species* which have an Aristotelian ring: 'I have hitherto spoken as if the variations . . . were due to chance. This, of course, is a wholly incorrect expression' (ch. 5). 'What limit can be put to this power [sc. natural selection] acting during long ages, and rigidly scrutinising the whole constitution, structure, and habits of each creature,—favouring the good and rejecting the bad?' (ch. 15). 'Natural selection acts solely by accumulating slight, successive, favourable variations . . . we can see why throughout nature the same general end is gained by an almost infinite diversity of means, for every peculiarity when once acquired is long inherited, and structures already modified in many different ways have to be adapted for the same general purpose' (ibid.). Aristotle had no theory of evolution of species, and there are important differences between Darwin's conception of the 'struggle for existence' (v. *Origin*, ch. 3) and Aristotle's conception of nature as form, but both are agreed that in accounting for the parts and movements of a thing we must consider what good they do. For some recent remarks on Aristotle's position *vis-à-vis* Darwin see Professor M. Grene, *Portrait of Aristotle*, pp. 145–8.

Having sketched the opposing case, Aristotle proceeds to develop his own in two or perhaps three main lines of argument. The first is pursued in **198b32–199a8** and **199b18–26**. The things Aristotle wishes to show are for something, are either for something or due to chance or coincidence (**199a3–4**). But that which is due to chance does not come about always or usually in the same way (cf. 196b10–20), and these things do (**198b35–6, 199b24–6**). They must, therefore, be for something.

If Aristotle were arguing that whatever is due to nature is for something, this argument would clearly be worthless, for it will be simply denied that all phenomena in the world of nature are either due to chance or for something. As observed above, however, Aristotle is discussing only a limited class of natural phenomena (the words in **198b34–5** 'The things mentioned, and all things which are due to nature' mean simply 'These things, like everything else due to nature')—the parts of living things (198b23, 28) and, in general, the class of things marked out in 196b17–22 as such as might be due to nature (sc. in the sense of form, see above, p. 106). There need be no circularity in defining them as things which might be due to nature, when the aim is to prove they are due to nature: the nature or function of a thing can be defined as suggested above, p. 102, and the things which might be due to nature are the things necessary or desirable for the performance of this function. Aristotle is, I think, right to say that these things seem to be either for the living thing and the performance of its function or the outcome of chance. What is questionable in his argument is whether they cannot be due to chance if they come about always or usually in the same way. It is true that we do not in ordinary speech call a thing the outcome of chance if it is usual, or say that it is by chance that a seed grows into a thing with roots and leaves, or a foetus into a thing with legs and teeth. But in this case the testimony of ordinary speech is not enough: we must go behind it, and inquire whether these processes which come about always or usually in the same way may not still be explicable only in the way in which a stool's landing on its feet is explicable. As Aristotle himself observes (198b18–21), the rain regularly falls in such a way as to make the crops grow, but we do not think it falls for that purpose.

The first line of argument, then, seems inconclusive. The second, which is not wholly separate from the third, appears in **199b13–18,** and perhaps **a8–12** and **b7–9.**

In **a9–10** Aristotle says that 'as things are done, so they are by nature such as to be'. He may mean: 'as are processes of manufacture, so are processes of nature.' In that case lines a8–12 belong to his third argument, that if art is for something so is nature, though he will here be simply asserting this, instead of trying to prove it as he should. Alternatively he may mean that if a process of production, whether natural or artificial, is for something, so is the product. That would be a better point, but then 'things are done for something' (a11) must mean 'natural processes are for something'; and this again is a *petitio principii.*

In **b7–9** and **13–18** Aristotle seems to be recurring to the point made in 194b29–30, that wherever a continuous change has a definite end in the sense of last thing, that end is also an end in the sense of thing for which. The growth of a living thing is such a process: living things do not come to be at random, but from definite kinds of seed (for an eloquent expansion of this theme see Lucretius i. 159–214); so growth is for the mature plant or animal and the performance of its function. This would make the assertion that natural processes are for something less bald and arbitrary; but as observed above, p. 101, the principle stated in 194b29–30 is not acceptable. Even if we can give grounds for relating processes in living things to the maturity of the living thing and not to its death, we shall still have a basis only for an 'als ob' teleology: processes will only be such as might be due to nature. Wieland welcomes this consequence, since he thinks that Aristotle's teleology is actually of this character; but apart from the fact that this in general seems untrue, Aristotle would have no grounds for restricting teleological explanation to living things. The full moon would be what its phases are for; or, still clearer, being at the centre of the universe would be what earth or its movement is for.

The second line of argument, then, seems as inconclusive as the first. The third appears in **199a12–b7, b26–33,** and is that if that which is in accordance with art is for something, so is that which is in accordance with nature (a17–18, cf. b30).

Aristotle clearly does not anticipate any dispute that things in accordance with art, i.e. artefacts and the movements of craftsmen, doctors, etc., *are* for something. In fact, however, a rigid upholder of mechanical determinism would deny this: he would say that the cobbler's hands move as they do, because of the forces to which the particles constituting his body are subject, and it is just by chance that there comes into being something such as to fit and protect a man's foot; human actions are analysable without remainder into the movements of particles in accordance with laws of physics. Although Aristotle does not offer us a counter to this suggestion, one might be derived from the consideration mentioned above, pp. 106–7, that we should perhaps never be able to notice anything, whether it is in fact for something or not, unless it seemed such as to be for something. If this is so, and we pick out and bring under a concept a pebble because it would do as a marble, a mountain because it would be good to ski down or hard to walk over, and so on, is it not necessary that we should sometimes make movements *for* a game of marbles, or skiing, or reaching a destination? Could a saw be noted as well adapted for cutting wood, if we never used it for cutting wood, but only observed that when it moves in a certain way, wood is divided—especially if we were never able to use the pieces thus separated for anything? An infant, perhaps, does start by noticing that a movement on the part of an object is followed by a glint or bang; but it is doubtful how far its intellectual development would proceed if it was not capable of repeating the movement for the sake of seeing the glint or hearing the bang. And if the movements of human beings are sometimes for something, 'the for something' is a cause of the things that result from these movements. Along these lines, then, it could be argued that that which is in accordance with art is for something. This argument, however (and the cruder one that we know by introspection that our actions are purposive), is not applicable to the processes of nature; can Aristotle show that they are still so like the processes of art that they too must be called for something? He offers several considerations.

In **199ᵃ15–17** he says that the practice of art is merely a continuation (e.g. medicine, agriculture) or copy (e.g. painting, choreography) of the action of nature, so if the former is for something, so should be the latter. Similarly in **ᵃ33–ᵇ7** he points out that mistakes occur in the practice of art, and monsters can be explained as analogous. These considerations, I think, carry little appreciable weight.

In **199ᵃ20–30**, Aristotle says that swallows and spiders do not act from knowledge or deliberation, but their movements when they make nests and webs are surely for something. If you say that nothing is for something unless it is done from deliberation, then you exclude much that is the exercise of art (199ᵇ28); thus a man exercising the art of the scribe does not deliberate how to spell (*E.N.* III 1112ᵇ2). And if the movements of spiders and swallows are sometimes for something, why not say that the roots of plants grow downward, not because they are made of a certain sort of stuff, i.e. earth, which necessarily moves downwards (cf. *De an.* II 415ᵇ29–416ᵃ1) but for nutrition (199ᵃ28–9)? To this it might be answered that we say the movements of swallows are for something because they seem to be directed by thought or at least perception, and deny that the behaviour of plants is for anything in so far as we deny to plants any kind of awareness.

In **199ᵃ12–15** there is the curious argument that if artefacts were natural objects, and conversely if natural objects were artefacts, they would come to be in just the way they do now. This argument seems to be continuous with the argument of **ᵃ8–12** discussed above, about changes being for the ends to which they lead, and hence is usually taken to be: if a house were a natural object its parts would be formed in the same temporal order in which they are formed now, first the foundations, then the walls, etc.; and conversely, if a man were an artefact, his parts would be formed in the same order in which they are now, first the heart and so on. Aristotle is certainly interested in the order in which the parts of living things are formed (see *De gen. an.* II 742ᵃ16–ᵇ17), and the usual interpretation may well be correct. If, however, we consider the *De gen. an.* II passage carefully, we may feel that Aristotle would not have thought that a house or ship would develop by nature precisely as it develops at the hands of the builder or shipwright (Empedocles with his neckless heads might think that, but Aristotle would more likely expect a house to grow like a mushroom, a ship to be hatched out of an egg); and that he is concerned rather with logical priority and posteriority and the subordination of means to ends (compare 199ᵃ15 with 742ᵇ7–8). In that case, his point will be that the roof would not be any less for protection or the rudder for steering if the house or ship were due to nature; and conversely the fins of a fish would not be any more for propelling it through the water if a fish were an artefact. The later passage in which he returns to the idea of nature producing what are now products of art, and where there is no mention of the order in which parts are formed, may be understood in the same way.

If this is Aristotle's argument, it has a certain quaint appeal. If we say that oars and ship-building are for something, have we any ground for denying that fins and the growth of fishes are for something, except that they are not due to thought, and if not, is that a sufficient reason? Are not natural processes, as Aristotle says, very like the case of a doctor's healing himself (**199b31**), especially, we might add, if his knowledge of medicine is so much a part of himself that he takes the right pill instinctively, without deliberating, as soon as he feels the onset of a cold?

Still by itself this argument cannot bear the weight of a full-bodied teleology, and our verdict may well be that Aristotle's defence of the thesis that some things due to nature are for something is inconclusive. I am not sure, however, that that is because the thesis is in fact indefensible. Aristotle is in effect trying to give an account of the difference we feel there to be between living things and the processes of life on the one hand, and inanimate nature on the other. It is not an absurd suggestion that living things are things to which we apply descriptions different in character from 'so much stuff of this sort shaped and arranged thus', or that the difference lies precisely in this, that to think of things under these other descriptions is to place them in the field of teleological explanation. And there is surely something in the argument that if what is due to art is for something, so is that which is due to nature. We cannot say that it is merely a matter of convenience that we think of rational human actions as explainable in terms of reasons, since convenience is itself a teleological notion; and it is hard to think it is due to chance that men have the parts they need to perform rational actions, unless we suppose that immaterial souls hover about waiting to pounce on suitably constituted bodies (cf. *De an.* II 414a22–4). Aristotle might have done well, however, to put more emphasis on the notion of awareness, and its connection with teleological explanation. A factor, after all, can explain a thing's behaviour by showing the good of it, only if the thing is in some way, however dimly, aware of that factor; if Aristotle wants to maintain that the roots of plants grow down for food (199a28–9), he ought to allow plants some sensitivity to wet and dry. And second, he might have emphasized that the validity of teleological explanation is an empirical issue. He is aware that under a description like 'Diares' son' an object cannot affect our sense-organs (*De an.* II 418a21–4). Since it is under such a description that it would provide a reason for action, the empirical question arises: can our movements be explained mechanically by action on our sense-organs? If so, pursuit and avoidance are epiphenomenal; if not, our movements are for something.

CHAPTER 9

In this chapter Aristotle tackles the second question raised in 198^b10–12, the role of necessity in natural things. The natural things he is concerned with do not, it seems, include the elements and their natural movements, but are those things only which it was argued in chapter 8 are for something. As for necessity, in general Aristotle recognizes three sorts of necessity (see *Met.* Δ 5), conditional necessity, unconditional necessity, and constraint. The last is not in question here. The theme of the chapter is that, of the two factors involved in living things, matter and form, the former is necessary and the latter is not, and the necessity is conditional. Matter must necessarily be present and undergo certain changes *if* the form, the dog or tree, is to exist or come to be. We might think that although the form is not necessitated by the matter in this way—there does not have to be a dog or tree *if* there is to be earthy or fiery stuff—still, it also is necessary in a way: if the right matter is affected in the right way, a dog or tree results with the same necessity with which air results when water is heated. However, Aristotle would doubtless reply that warm wet stuff can remain warm wet stuff without being converted into an animal; if it is converted into an animal, that is through the agency of a parent or because of the life already imparted to it by a parent (cf. *De gen. an.* II 735^a12–13, 18–22). (In fact, if an animal is a thing the behaviour of which cannot be explained in accordance with mechanical laws of physics, then the coming into being of an animal cannot be predicted in accordance with those laws as the outcome of action on the matter which is converted into its body.)

The argument of the chapter is fairly straightforward, and most of the ideas behind it have already been discussed. A couple of points, however, may bear comment. First, the analogy with mathematics (**200^a15–30**). We are told that the properties of geometrical figures are due to the formal factor, and by the formal factor Aristotle understands the nature of the simple elements into which a figure can be resolved, straight lines, curves, etc. (198^a16–18). This seems arbitrary: we might think that lines stand to triangles as matter, and that explaining the properties of a triangle by the nature of straight lines is like explaining the movement of a stone by the nature of earth. Anyhow, Aristotle says that explanation in mathematics is like explanation by ends in natural science, but the reverse: like, because the properties of a triangle are rendered necessary by the nature of straight lines—a triangle must have angles together equal to two right angles, *if* we are to have straight lines (not, perhaps, a very Euclidean way of putting the matter); the reverse, because in natural things the factor which necessitates, the mature living thing or fully formed organ, is the last thing to come to be, whilst in geometry the factor which necessitates, the definition of a straight line, is the starting-point. This is best regarded as mere architectonic. In general (e.g. *De gen. an.* II 734a30–1) Aristotle thinks that nature in the sense of form must come first.

Second, it is suggested in **200b4–8** that there are parts of the account which stand to it as matter. Elsewhere Aristotle suggests that in a definition by genus and differentia, the genus stands to the differentia as matter to form (*Met. Z* 1038a6–8), but what he probably has in mind here is that natural things are like snubness (see above, pp. 95–6): the matter must enter into the account of a living thing or organic part, and play the role there which noses play in the account of snubness.

APPENDIX

Did Aristotle Believe in Prime Matter?

In the commentary I express scepticism about the traditional view that Aristotle believed there is a single, eternal, and completely indeterminate substratum to all physical change, called prime matter. Some remarks by Zeller in his *Aristotle* will illustrate this view: 'Becoming in general . . . presupposes a substratum whose essence it is to be pure possibility (p. 342) . . . presupposes some Being . . . which underlies as their subject the changing properties and conditions, and maintains itself in them (p. 344) . . . This substratum cannot itself ever have a commencement; and since everything which perishes resolves itself finally into the same substratum, it is imperishable also (p. 345) . . . If we abstract entirely from everything which is a product of becoming . . . then we shall have pure Matter without any determination by Form. This will be that which is nothing, but can become everything—the Subject, namely, or substratum to which no one of all the thinkable predicates belongs, but which precisely on that account is receptive of them all . . . This pure matter . . . Aristotle calls *πρώτη ὕλη* ["prime matter"]' (pp. 247–8). I here append reasons for rejecting this account of Aristotle's teaching, and also some suggestions about how it arose.

(1) We may start with the phrase *prōtē hulē*, 'primary matter', itself. By this is traditionally understood the *ultimate* substratum of change. Now this expression does occur, though not often, in Aristotle. Bonitz lists the following places: *Phys.* II 193a29, *De gen. an.* I 729a32, *Met. Δ* 1015a7–10, *H* 1044a23, *Θ* 1049a24–7. To this we may add *Met. Δ* 1014b32, 1017a5–6, and (passages where Aristotle speaks of a *prōton hupokeimenon* or *enuparchon*, 'primary underlying thing' or 'constituent') *Phys.* I 192a31, II 193a10.

In 193a10, 193a29, 1014b32, and (cf. 1016a19–24) 1017a5, primary clearly means 'proximate'. Similarly, I think (see above, p. 83) in 192a31.

In 729^{a}32 the meaning is obscure, and no one would try to base anything on that uncertain text. In 1044^{a}23 also the meaning is doubtful. If the reading of the MSS., which have a reference to 'primary matter' in the preceding line ^{a}18, is retained, it means 'proximate' in both places; if, with Ross, we delete the reference to primary matter in ^{a}18, it may mean ultimate, but if so it probably means ultimate in the same sense as in 1049^{a}24–7.

In 1049^{a}24–7, 'primary' means 'ultimate', but the matter referred to there is determinate, and in any case Aristotle does not commit himself to it. 'If', he says, 'there is some primary thing which is not called "that-y" (*ekeininon*) in relation to anything else, that is primary matter. Thus if earth is airy, and air is not fire but fiery, fire will be primary matter which is not a this thing here.' Three points about this passage. First, Aristotle is obviously speaking of the possibility of an ultimate determinate matter, e.g. fire, to be discovered by the student of nature, not of an ultimate indeterminate matter, to be discovered by conceptual analysis. Second, Aristotle in fact rejects this possibility when he comes to consider it in its proper place, *De gen. et cor.* II 332^{a}4–20. Third, my translation follows Ross's text, according to which the final phrase is *ou tode ti ousa* 'not being a this thing here'; earlier texts read *hōs tode ti kai ousia*, 'as a this thing here and reality'. This reading, though, it seems to me, in view of the lines which follow, certainly wrong, would tend to encourage the traditional view that Aristotle believed in primary matter.

Finally, we have 1015^{a}7–10, which is generally agreed to be the most important text for the phrase 'primary matter'. It runs as follows: 'Nature is the primary matter—and that may be either of two things, that which is primary in relation to the thing, or that which is primary in general; thus in the case of works of bronze, bronze is primary relative to them, but water, perhaps, is primary in general, if all the things which can be melted are water.' It seems to me that 'primary' here means 'proximate'; and that two things, bronze and water, may either of them be called the primary matter, because that of which we are trying to specify the matter may be described in specific terms, as bronze artefacts, or in generic terms, as things which can be melted. See 1023^{a}26–9: 'A thing is said to be out of something in one sense, it is out of it as its matter; and this in two ways, according to the primary genus or the last species; thus there is a way in which all things which can be melted are out of water, and a way in which the statue is out of bronze'; and compare *Phys.* II 195^{b}22–7: 'We should always look for the topmost [i.e. proximate] cause . . . we should look for kinds of cause for kinds of thing and particular causes for particular things.' Even if this interpretation of 1015^{a}7–10 meets with reservations, it is clear that here, where if anywhere Aristotle should mention his indeterminate universal substratum as a possible reference of the expression 'primary matter', he does not do so. I conclude, therefore, that whether or not he believed in such a substratum, 'primary matter' is not his expression for it.

(2) In *De gen. et cor.* (I 317^{b}13, II 329^{a}27) and elsewhere, Aristotle refers us for a full discussion of the notions of form and matter to *Phys.* I. The crucial passages in *Phys.* I are 191^{a}8–12 and 192^{a}25–34, and we have seen that these provide no evidence that Aristotle believed in prime matter, whilst the general tendency of *Phys.* I is precisely to show us how to avoid the need to posit a single universal substratum. Alongside the evidence of *Phys.* I we may put two or three other passages. In *De part. an.* II 649^{a}20–1 Aristotle entertains the possibility that smoke or charcoal stands as underlying thing to fire. In *Phys.* IV 213^{a}2–4 he says: 'Water is the matter of air, and air is as a sort of realization of water. For water is air in possibility, and air is water in possibility in another way.' Such passages need explaining away by anyone who thinks that prime matter is what stands to the elements as matter.

Again, *Met. H* 1042^{a}32–^{b}3: 'That the matter also is reality (*ousia*), is clear: in all opposed changes there is something underlying the change, for instance in a change of place that which is now here and now elsewhere, and in a change in respect of increase, that which is now so great and now smaller or greater, and in an alteration, that which is now healthy and now sick; and similarly in respect of reality, that which is now in coming to be and now in passing away, and now underlying as a this thing here, and now underlying in respect of a lack.' This passage brings out well the difference, already noted above, p. 77, between Aristotle's uses of the expressions 'underlying thing' and 'thing which remains'. The underlying thing is the *terminus a quo* of a change under whatever description, and here (^{b}3) includes the lack, which certainly does not remain; hence Aristotle's insistence that there is always an underlying thing is no evidence that he thought there is always something which remains. And we also find Aristotle saying that the underlying thing in cases of coming into being and passing away, when or in so far as it is *not* a lack or opposed, is a 'this thing here', i.e. something like a seed or an animal; it is incredible that he should have called the underlying thing in such cases a 'this thing here' if he had conceived it as prime matter. The words 'that which is now in coming to be and now in passing away' are not absolutely clear. Aristotle might mean that the underlying thing is, in cases of coming to be, that which a reality comes to be out of, and, in cases of passing away, that which it passes away into; or he might mean that it is that which is sometimes in process of coming to be (sc. when a reality is passing away), and sometimes in process of passing away (sc. when a reality is coming to be). But however we interpret the words, the suggestion of the passage is surely that in cases of coming to be and passing away there is always a definite underlying thing we can get hold of, not that there is an underlying thing which evades our grasp because it is indefinite and imperceptible.

(3) The principal passages on which the traditional view relies are in *De gen. et cor.* II. I quote the best-known one in the translation of Joachim, who is a strong supporter of the traditional view. 'Our own doctrine is that although there is a matter of the perceptible bodies (a matter out of which the so-called "elements" come to be), it has no separate existence, but is always bound up with a contrariety. A more precise account of these presuppositions has been given in another work' [Joachim has here a footnote: 'Cf. *Physics A*. 6–9, where πρώτη ὕλη and "the contrariety" (εἶδος and στέρησις) are accurately defined and distinguished as presuppositions of γένεσις']: 'we must, however, give a detailed explanation of the primary bodies as well, since they too are similarly derived from the matter. We must reckon as an "originative source" and as "primary" the matter which underlies, though it is inseparable from, the contrary qualities: for "the hot" is not matter for "the cold" nor "the cold" for "the hot", but the *substratum* is matter for them both. We therefore have to recognize three "originative sources": *firstly* that which is potentially perceptible body, *secondly* the contrarieties (I mean, e.g., heat and cold), and *thirdly* Fire, Water, and the like' (329ᵃ24–35). Another passage which might seem to tell in favour of prime matter is 332ᵃ34–ᵇ1: 'If there is a single

opposition in respect of which they change, the elements must be two; for the matter, the intermediate, is imperceptible and incapable of separate existence.' Somewhat of a piece with these passages is *Met.* Λ 1070^{b}10–13: 'In one way the elements of everything are the same, and in another not. Thus perhaps in the case of perceptible bodies, the element as form is the hot, and the cold is an element in another way, as the lack, and the matter is that primary thing which is of itself these in possibility.'

These passages as they stand may appear to be quite good evidence that Aristotle believed in prime matter. We ought, however to consider their contexts, and the general line of argument running up to *De gen. et cor.* II.

In *De caelo* III. 6 Aristotle argues that the elements are not eternal, but come into existence and pass out of existence, and do so by changing into one another. One of his arguments is as follows: 'It is not possible that the elements should come to be out of any body. For if they do, it will follow that there is some body prior to them. If this has weight, it will be one of the elements; if it has no weight, it will be unchangeable and an object of mathematics' —in which case nothing can come to be from it. '... If, then, the elements can come to be neither from that which is incorporeal, nor from any other body, it remains that they come to be from one another' (305^{a}22–32).

This hardly prepares us for the introduction of a universal indeterminate substratum, and neither does *De caelo* IV 312^{b}20 ff., where Aristotle argues for there being as many kinds of matter as there are kinds of body, and against 'a single matter of all things, such as void or plenum or extension or triangles'.

De gen. et cor. I opens with the observation that those who posit only one kind of matter, or make everything come to be out of one thing, must say that what we call cases of coming into existence and ceasing to exist are really cases of alteration (314^{a}8–13, ^{b}1–5). In chapter 3 he says that there are difficulties about supposing that realities come into being. If a reality comes into being, it must do so out of something which is a reality in possibility: if this has affections like size and place in actuality, but is not a reality in actuality, then affections will be capable of separate existence; if it has nothing in actuality, then something which is in actuality nothing will be capable of separate existence (317^{b}23–33). But, he says, these difficulties can be avoided if we suppose that the coming to be of one reality is always the ceasing to be of another and vice versa, i.e. if we suppose that realities change into one another (318ª23–7). In chapter 4 he goes on to distinguish coming into existence and alteration, and offers the criterion we saw above, p. 75; what we would expect him to say, then, is that a change of one reality or element into another can be a genuine case of coming into existence, so long as no underlying thing remains throughout the change.

In *De gen. et cor.* II, immediately before our first crucial passage, 329ª24–35, Aristotle criticizes 'those who posit one matter over and above those mentioned', sc. the elements, 'and that corporeal and separable': 'it is impossible that this body should be without perceptible opposites' (329ª8–11). He also attacks Plato's account in the *Timaeus*: 'He does not speak clearly about the all-receptive, whether it is separate from the elements. Nor does he use it, saying that there is some prior thing underlying the elements, like gold in relation to works of gold. And indeed, even this is not well said, when it is spoken of thus. We speak so of things which undergo alteration, but when a thing comes to be or passes away, it is impossible to call it that out of which it has come to be. Yet he says that it is far truest to call each thing gold.' Aristotle's immediate point here may be the one noticed above, p. 75, that (e.g.) a gold ring should be called not gold but golden; but there is no suggestion that Plato would have escaped censure if he had said that material objects generally were not space but space-en. Similarly just before our second crucial passage, 332ª34–ᵇ1, Aristotle argues that there cannot be a single matter for everything which is (*a*) one of the elements—since then all change would be alteration (332ª6–13)—or (*b*) 'something over and above them, such as some sort of intermediate between air and water or air and fire'—since this would have to be 'either any one of them indifferently, or nothing at all' (ª20–6).

Finally, Aristotle's promise in 329ª24–35 that he will explain the transformation of the elements is redeemed in the following three chapters. In those chapters there is no mention of a wholly indeterminate substratum; the nearest Aristotle comes to such a thing is in speaking of certain bodies 'simpler' than the elements, which should be called not fire but 'fiery-in-form', not air but 'airy-in-form' (330ᵇ21–5), and which, it seems, come to be out of, and pass away into, one another (331ª7–8). Further, summing up his theory in chapter 9, Aristotle says that the principles are three in number, matter, form, and source of change, and 'the cause as matter of things which come to be is that which is able to be and not be' (335^a28–33), which is intelligible if the material principle is that which ceases to be when something comes to be out of it, but not so intelligible if it is eternal prime matter.

Clearly, then, the general atmosphere of the *De caelo* and *De gen. et cor.* is not propitious to the introduction of prime matter. Aristotle has ruled out the following kinds of universal substratum: incorporeal, corporeal, void, plenum, extension, triangles, objects of mathematics, Timaean space, some one of the elements, something over and above the elements. It might be urged that his exclusion of universal perceptible substrata (cf. 319^b10–16) implies an acceptance of a universal *imperceptible* substratum. It is hard, however, to see how an imperceptible substratum could be corporeal, since the qualities perceptible by touch—hot, cold, wet, and dry—are the differentiating features of body as such (v. *De an.* II 423^b26–9). Again, it is sometimes said that prime matter escapes the objections to other substrata because it is not capable of separate existence; but against this we should set the explicit statement of 317^b28–9, that if realities came to be out of something which was 'all things in possibility, it follows that a non-being of this sort is capable of separate existence'. Of course, if we define prime matter as that universal substratum, whatever it may be, which Aristotle does not explicitly exclude, we can then say that he does not exclude it; we should not, however, be surprised if it then turns out to be very odd stuff.

Let us see, then, whether there is any alternative to taking the passages cited as asserting the existence of prime matter. If we bear it in mind that Aristotle makes careful use, in the *De gen. et cor.* as elsewhere (see above, p. 77), of the two distinct terms 'underlies' and 'remains', the former extending more widely than the latter, we might paraphrase 329^a24–7 as follows: 'We agree with Plato and others that nothing comes to be out of nothing. There is always something which underlies, in the sense indicated in *Phys.* I. 7, something out of which the new element, or whatever it is, comes to be. This, however, is not something separate but always characterized by hot or cold, wet or dry'. (The point is not that it cannot exist on its own, but that it is not something separate from the elements or from their definitive qualities—

unlike, e.g., bronze, which constitutes spheres and cubes, but it is not in itself characterized by straight or curved.) It is always another element, with one of a pair of opposed qualities. It is thus, like any other underlying thing, two things in account. That out of which a cold element arises may be hot, and under that description is one of the opposed principles; but it is not under that description that it is the material factor, but under the description "perceptible body in possibility" or "cold body in possibility".'

The other two passages may be understood in a similar way. 332ª34–ᵇ1 comes in a chapter in which Aristotle is most naturally understood as trying to show that there are exactly four types of ultimate substratum (all of them definite). Whilst the precise sense of the crucial words is obscure, as Joachim himself admits, the point Aristotle wishes to make can only be that if a change is between one pair of opposites, e.g. hot and cold, then even if there is an underlying thing or intermediate, it is not a third factor over and above the two elements which are the termini of the change, the hot one and the cold one (cf. ª24–6). 1070ᵇ10–13 comes in a passage where Aristotle is arguing that the material factor in a change from one element to another is not the same as, but only analogous to, the material factor in an alteration. I do not think this need be taken to imply that the material factor in a change from, say, earth to fire, is not just analogous to but the same as that in a change from air to water. (For different levels of sameness by analogy see *Met.* Θ 1048ª35–ᵇ9.) Aristotle might well think that that which is 'of itself in possibility hot' is earth or water, that that which is 'of itself in possibility wet' is earth or fire, etc. The laconic parenthesis 'but different things are the principles of different things' (1070ᵇ17) gives some support to this interpretation, though it might be taken to refer only to the coming to be of composites like flesh and bone (cf. ᵇ15), and not to the coming to be of elements.

(4) Another passage in which Aristotle has been thought to commit himself to prime matter, is in *Met. Z.* 3. I offer a translation of it in its context. 'The underlying thing is that of which the other things are said, and which is not itself said of anything else. Let us then begin by getting that clear. For the best claim to be called reality (*ousia*) seems to belong to the primary underlying thing. In one way the matter is called such, in another the form, in a

third that which consists of these. (By the matter I mean, e.g., the bronze, by the form, the shape of the idea, by that which consists of these, the composite statue.) So if the form is prior to the matter and more of a thing which is, it will be prior also to that which consists of both on the same account. We have now said in outline what reality is: it is that which is not of an underlying thing, but which the other things are of. But we must not leave it at that: it is not enough; what has just been said is unclear, and further, the matter comes to be reality. For if it is not, it is hard to see what else can be. For when the other things are cut away from round about, nothing at all appears remaining. For the other things are affections, operations, and powers of bodies, and length, breadth, and depth are not realities but quantities (an amount is not a reality), but it is rather that primary thing to which these belong which is a reality. But when length, breadth, and depth have been stripped away, we see nothing left, unless there is something which is what these delimit; so that to anyone looking at the matter thus, it must appear that matter is the sole reality. And by matter I mean that which' (or: 'And I mean a matter which') 'in itself is called neither a definite thing, nor a definite amount, nor anything else by which that which is is defined. For there is something of which each of these things is predicated, which is different in being from any of the predications (for the other things are predicated of a reality, and that of the matter): so that the last thing is in itself neither a definite thing nor a definite amount nor anything else. Nor the denials, for these too will belong by virtue of concurrence. Those, then, who speculate along these lines will find that reality turns out to be the matter. But that is impossible. For to be separable and a this thing here seem to belong to realities above all. Hence the form and that which consists of both seem to be reality rather than the matter' ($1028^{b}36$–$1029^{a}30$). Bonitz ($785^{a}25$ ff.) quotes $^{a}20$–1, 'in itself is called neither a definite thing nor a definite amount nor anything else by which that which is is defined', as Aristotle's formal definition of matter.

The general sense of the passage seems to be as follows. The things which have the best claim to be called realities are dogs, houses, and the like, for sizes, shapes, etc. are the sizes, shapes, etc. of such things, and dogs and houses are not in the same way the dogs and houses of anything further. That, however, to which we apply the expression 'a dog' is in one aspect matter, in another form, in another that which consists of the two. In which aspect has it the best claim to be called a reality? At first we might think, in its material aspect. It is a dog in the sense of a quantity of flesh and bone, a statue in the sense of a quantity of bronze, which other things are said *of*. (The people who take this line may be identified with some confidence as the people who say that the nature and reality of things is their proximate matter, *Phys.* II 193ᵃ9–28.) Against this line, however, Aristotle argues that if we take it to its logical conclusion we shall be awarding the title of reality to something completely indeterminate, which is absurd.

So much is fairly clear. What is not clear is whether in ᵃ10–26 Aristotle is saying 'There is indeed a completely indeterminate substratum to everything, but it cannot be called reality', or 'If we say that bronze has more claim to the title of reality than what it constitutes, we shall then be forced to posit some completely indeterminate matter'. The second interpretation, however, is preferable. On the second interpretation, lines ᵃ10–26 are all a statement of an opponent's line of thought; and they are remarkably similar in feel to *Met. B* 1001ᵇ26–1002ᵃ14, a passage in which Aristotle is certainly sketching a line of thought he rejects. And further, the second interpretation yields the better argument. If we say (as Aristotle himself does) that the bronze is less of a reality than what it constitutes (that *this* bronze is the bronze of *this statue or sphere*), then we ought surely to say that the length, breadth, and depth are the length, breadth, and depth of the statue: 'what these delimit' will be the statue, not some indeterminate stuff. It is only if we deny that the formless is logically parasitic on the formed, that we can introduce prime matter at all, at least by the argument sketched here.

(5) There is also a passage we should consider in *Phys.* IV: 'We say that there is a single matter for the opposites, for hot and cold and the other natural oppositions, and what is in actuality comes to be from what is in possibility, and the matter is not separable, but its being is different, and it is one in number, if it so happens, for colour and for hot and cold. And there is matter also for body, the same for large and small. That is clear. For when air comes to be out of water, the same matter does not come to be through taking in something else, but what it was in possibility, it comes to be in actuality' (or: 'what was in possibility comes to be in

actuality') 'and it is the same when water comes to be out of air. . . . Similarly if a large amount of air comes to be in a smaller volume, or a smaller in a larger . . . For just as the same comes to be hot from cold and cold from hot, because it was in possibility, so it becomes hotter from hot, without anything coming to be hot in the matter, which was not hot when the thing was less hot . . . So that the largeness and smallness of perceptible bulk do not extend because something is added to the matter, but because the matter is in possibility both. So the dense and the rare are the same, and there is a single matter for them' ($217^{a}21–^{b}11$).

This looks as much like a commitment to prime matter as anything in the *De gen. et cor.* Once again, however, we must consider the context. Aristotle is arguing against the real existence of void. His adversaries, as he represents them, say that change of size, such as occurs when water turns to air, or when a tree grows, can be explained only on the assumption that there is void in bodies. When water changes to air, it expands because void comes into it; when a tree grows, the surrounding air minimally contracts, doubtless because void goes out of it. On this view, as Aristotle sees it, there will be two material factors when water changes to air, water and void. Aristotle's own suggestion is that such changes should be explained in the same way as changes in temperature. When a thing gets hotter, we do not suppose that there is new hot stuff in it, but simply that what was less hot is now more hot. Similarly when water turns to air, we need not suppose that new extended stuff, void, has come along, but simply that what was small and cold has turned into something warm and large in volume. The debate may seem to us a little artificial; but the point for our present purpose is that when Aristotle emphasizes that there is a single matter, he is denying that in changes of size things arise out of two things, matter and void; he is not considering whether the kinds of basic matter are one or several. And it is not reasonable to argue that, because Aristotle denies things are constituted out of a completely featureless void, he must believe they are constituted out of a completely featureless matter.

(6) The passages we have now considered are, I believe, all the evidence there is in the Aristotelian corpus that Aristotle believed in prime matter; and it seems to me that they will not carry the weight of the traditional view, that the traditional view must

be held to be at the best not proven. It derives, however, some support from a feeling, more widely held than admitted, that Aristotle does in fact need something like prime matter, that the positing of a universal substratum is in fact a conceptual necessity.

If we feel that all natural change is the transformation of something which remains throughout the change, it is probably because, like the Presocratics, we fear that otherwise we shall have the impossibility that things come into being out of, and pass away into, nothing. Be that as it may, there is a formidable difficulty about saying that anything remains throughout all change. A thing can be said to remain, or be the same, only under some description. To say that when air changes to water there is some *matter* which had the form of air and comes to have the form of water, is not enough: we must specify what the matter is. We cannot claim that that which was air is the same as that which is water, without saying the same what. And 'matter' (to say nothing of 'bit of prime matter') is not an adequate description.

The idea, however, that if there is nothing which remains throughout a change, then things come to be out of or pass away into nothing, is mistaken. Between alteration on the one hand, and creation and annihilation on the other, there is a third possibility. If you have a glass jar from within which you have removed the air and everything else you can find; and you see a frog suddenly appear in it; then you might call that coming to be out of nothing. If you see a man sitting in a chair, and suddenly he has vanished irretrievably, and in his place is a pile of books which have never been seen or heard of before, you might be tempted to say that the man has passed away into nothing and, by a strange chance, the books have come into being out of nothing in the same place. But when the passing away of one thing is always and intelligibly attended by the coming to be of another, for instance when wood passes away in smoke and flames, or a saucer of water passes away and the air is refreshed, then we do not say that the first thing has passed away into nothing, but into the second, and we say that the second has come into being, not out of nothing, but out of the first. Yet we cannot say that there is something which remained throughout and underwent these transformations, unless we can find some description under which this thing can be identified throughout.

It may be added that we can often find such a description if we look, not for something which constituted first one term of the change and then the other, but for something which first one term and then the other constituted. When Midas touches his table, wood passes away, gold comes to be, and the table remains the same. We might say that this case could be thought of in two ways: it is an alteration on the part of the table, for the table remains but changes in respect of colour, weight, hardness, etc. But it is a coming to be of gold and a passing away of wood, since when the table undergoes these qualitative changes, the description 'wood' no longer applies to it in its material aspect, and the description 'gold' comes to apply. If this example seems too miraculous, there are said to be streams whose waters have a petrifying power: a table immersed in one long enough would be the same table but change in respect of its matter. When, of course, the thing constituted remains throughout a change in respect of matter, the change cannot be called a coming to be without qualification, since the *terminus ad quem* is parasitic on, is the matter of, that which remains.

It might also be felt (cf. G. E. M. Anscombe and P. T. Geach, *Three philosophers*, pp. 46, 72–3) that prime matter is needed as an ultimate principle of multiplicity and individuation: two pennies are two because they are made of two different bits of it. Now in the first place, a determinate matter will do as well as an indeterminate for this purpose. And second, while matter is needed as a principle of multiplicity in the obvious sense that if you want many pennies you need much copper, it is not a satisfactory principle of individuation. We would do better to say that things like dogs and pennies are 'precisely what are' individuals and countable units; so that pennies are not two because they are made of two pieces of copper, but pieces of copper are two because they make two pennies.

(7) Having said why I do not accept the traditional view, I now offer some suggestions as to how it arose.

The language of the traditional descriptions of prime matter originates in the *Timaeus*: 'Now the argument looks as if it forces us to try to clarify this dim and difficult sort of thing in words' (*Tim.* 49 a 3–4). 'It receives all things and never anywhere in any way takes any form like any of the things which come in; for it lies there, by nature the matrix for everything, and is changed and variously shaped by the things which come in' (50 b 8–c 3). 'That which is to receive all kinds into itself, must be outside all forms' (50 e 4–5). 'Hence the mother and receptacle of the visible and in general perceptible offspring should not be called earth or air or fire or water, or any of the things which arise from these or from which these arise; but if we call it an invisible and shapeless kind of thing, all receptive, but partaking in a most baffling way of the intelligible, and very hard to get hold of, we shall not lie' (51 a 4–b 2). 'And the third kind of thing is space, which always is, admits of no passing away, provides a seat for all things which come to be, and itself is to be grasped without perception by a sort of bastard reasoning, a thing hardly credible, which we see in a dream' (52 a 8–b 3).

These passages are much better evidence for a belief in prime matter than anything which can be found in Aristotle, and there can be little doubt that we have here the origin of the way in which prime matter is traditionally described. That we have the origin of the notion of prime matter itself would not be an untenable thesis; but in fact it seems to me that the notion of prime matter was reached by putting together Plato's language with Aristotle's concept of a material factor, by adapting Aristotle's underlying thing so that the Timaean account would fit it. Aristotle must bear some responsibility for this development, since he certainly represents Plato as trying in the *Timaeus* to characterize the underlying nature. His qualifications—that Plato did not go deep enough (*Phys.* I 191^{b}35–6), that the *Timaeus* is ambiguous (*De gen. et cor.* II 329^{a}13–14), etc.—were ignored by later thinkers, and we soon find it taken for granted that Aristotelian matter and the Timaean receptacle are the same thing; and that to which Plato's and Aristotle's descriptions are both more or less applicable is prime matter as traditionally conceived.

Prime matter seems to make its first genuine appearance among the Stoics, who added the doctrine, of which no trace is to be found in either Plato or Aristotle, that prime matter is *ousia*, reality, or (the word now becomes appropriate) substance. See, for instance, Diogenes Laertius vii. 150: 'They say that prime matter is the substance of all things which are: so Chrysippus in his *Physics* Bk. I, and Zeno.'

Prime matter as a conflation of Platonic and Aristotelian ideas is well established in the syncretist philosophy of the first century B.C. and the early centuries A.D. In a passage from Ocellus, cited by Miss C. J. de Vogel (*Greek Philosophy* III, 1280 b), the Platonic terms *ekmageion* and *pandeches* ('matrix', 'all-receptive') are worked in side by side with a quotation from the *De gen. et cor.* (II 329^a32–$^b3$). In the well-known description of matter in Albinus viii. 2: 'He calls this then matrix and all-receptive and nurse and mother and space and underlying thing, to be grasped without perception and by a bastard reasoning', every term is taken from the *Timaeus* except the all-important *hupokeimenon*, 'underlying thing'. And whilst the Timaean account is being understood as a description of matter, Aristotelian matter is being understood as wholly indeterminate. In the fragments of Nicolaus Damascenus' influential summary of Aristotle (? late first century B.C.) we find: 'The other matter, which is supreme, is wholly unspecified and without form', 'Substance is also said to be the ultimate substratum of everything and receptive of all forms. And that is matter' (fragments 9. 3 and 23. 5, translated by H. J. Drossaart Lulofs). The first passage is clearly a misinterpretation of *Met.* Δ 1015^a7 ff., and the second of 1017^b23–4.

One further thing is needed for prime matter to cement itself in European thinking: it must become acceptable to Christian theology. The credit for making it so probably belongs to pre-Christian Hellenizing Jews, who identified the 'invisible and shapeless earth' created along with Heaven in the first verse of Genesis with Timaean matter. So Calcidius 276–8, no doubt on the authority of Origen (see J. C. M. van Winden, *Calcidius on Matter*, ad loc.). The Christian fathers accepted the identification with enthusiam; as particularly significant for western thought, we may notice that Augustine adopts it without question, *De Genesi contra Manichaeos* i. 5–7, *de Genesi ad litteram* i. 14–15. (See also *Contra Faustum Manichaeum* xx. 14: 'The Greeks define *hulē*, in their discussions of nature, as a sort of matter of things, absolutely unformed, but capable of receiving all corporeal forms; which is recognized somehow or other in the changeableness of bodies, for by itself it can neither be perceived nor understood.')

The position, then, in the early centuries A.D. is that everyone believes in prime matter. It is found in Plato, Aristotle, the Stoa, and even in the opening words of the Bible. No discrepancies between Plato and Aristotle are detected. Hippolytus, in *Contra haereses*

i. 19 (Migne xvi. 3041 ff.), puts at the top of the list of Platonic doctrines: 'Plato says that the principles of the universe are God, matter, and paradigm'; Aristotle, he continues (ibid. 20), posits 'as principles of all things, substance (*ousia*) and accident; the substance is one single substance underlying everything'; it is thus hardly surprising that he concludes that Aristotle 'is pretty well completely in agreement with Plato' (σχεδὸν τὰ πλεῖστα σύμφωνος, ibid.). Similarly Simplicius in *Phys.* I, 191ᵃ5 declares 'There is hardly any disagreement at all between Plato's and Aristotle's accounts of the elements' i.e. of matter (ed. Diels 225. 17–19); he then actually goes on to explain what Aristotle must have thought about matter by quoting the *Timaeus* (226. 2–5), and interprets *kat' analogian* ('by analogy') in 191ᵃ8 as Aristotle's fancy name for what Plato calls 'bastard reasoning' (226. 25–227. 18).

This state of opinion gets fossilized in Calcidius' commentary on the *Timaeus*, a work the historical importance of which can hardly be exaggerated. In the Latin-speaking West it was almost the sole and easily the fullest source for ancient metaphysics until the twelfth century. Besides taking it for granted that Aristotle is a Platonist (see van Winden, op. cit., p. 144), and using large chunks of *Phys.* I to elucidate the *Timaeus*, Calcidius offers a couple of considerations in support of prime matter which were to have a long innings in the history of philosophy: that the properties of a thing, its size, shape, and so on, need something to keep them together (303), and that unless there is a substratum which remains throughout a change, from one element to another, we shall have properties hanging unsupported (317–18). (This does not, of course, follow: when water changes to air, there need be no moment at which the cold is the cold neither of the water nor of the air, just as when a body comes to rest, there need be no moment at which the body is neither in motion nor at rest.)

Calcidius' version of the views of Plato and Aristotle remained unchecked for some eight centuries, and it is thus not surprising that prime matter became an integral part of western philosophical thinking. In Calcidius, however (as in inferior sources like Macrobius), matter is represented as a Platonic discovery: how did it come to be fathered on Aristotle? I think almost by accident. As new texts of Aristotle arrived, and his exponents scored notable academic successes, the name of Plato ceased to carry much weight, and any doctrine which anyone wished to commend had to be backed by the authority of the Philosopher *par excellence*. By the end of the thirteenth century orthodox philosophy is Aristotelian. The process can be observed if we compare the *Summa philosophiae* of pseudo-Grosseteste, which represents the older Augustinian tradition, with Aquinas. In the essay on matter (tract. iv), pseudo-Grosseteste treats Plato and Aristotle as equal authorities, in broad if not complete agreement, and cites both freely. Aquinas certainly believed that Plato had a theory of matter (*S.T.* i. 66, art. 2, cf. *in Phys.* I. 133), but in the *De principiis naturae*, written, perhaps, before the *Summa philosophiae*, but representing the thought of a new generation, he appeals throughout to Aristotle without mentioning Plato. It does not follow that he approached Aristotle with a fresh and open mind. The *De princ. nat.* is on the whole an accurate and clear summary of the doctrine of *Phys.* I–II, but in § 5 Aquinas says: 'And since in coming to be the matter or underlying thing (*subjectum*) remains, but the lack (*privatio*) does not, nor does that which consists of the matter and the lack, it follows that that matter which does not bring in the lack is permanent, whilst that which does is transient.' As was said above, this can be shown to be Aristotelian doctrine only when the coming to be is a case of alteration, of something's coming to be something, not when it is a case of something's coming simply into existence.

SELECT BIBLIOGRAPHY

The following is a list of the principal works cited in the introduction, commentary, and appendix:

H. Bonitz, *Index Aristotelicus*, Berlin, 1870.

H. Cherniss, *Aristotle's criticism of Plato and the Academy*, New York, 1962.

C. J. de Vogel, *Greek Philosophy* III, Leiden, 1964.

DK = H. Diels and W. Kranz, *Die Fragmente der Vorsokratiker*, Dublin/Zürich, 1966.

D. E. Gershenston and D. A. Greenberg, 'Aristotle confronts the Eleatics', *Phronesis* 1962, 137–51.

M. Grene, *A portrait of Aristotle*, London, 1963.

R. P. Hardie and R. K. Gaye, *Physica*, Works of Aristotle translated into English, ed. Ross, vol. ii, Oxford, 1930.

M. Hesse, 'Aristotle's logic of analogy', *Philosophical Quarterly* 1965, 328–40.

H. H. Joachim, *De generatione et corruptione*, Works of Aristotle translated into English, ed. Ross, vol. ii, Oxford, 1930.

A. P. D. Mourelatos, 'Aristotle's "powers" and modern empiricism', *Ratio* 1967, 97–104.

G. E. L. Owen, 'Aristotle on the snares of ontology', *New essays on Plato and Aristotle*, ed. R. Bambrough, London, 1965, pp. 69–95.

—— 'Logic and metaphysics in some early works of Aristotle', *Plato and Aristotle in the mid-fourth century*, edd. I. Düring and G. E. L. Owen, Göteborg, 1960, pp. 163–90.

Philoponus, *In Physicorum octo libros commentaria*, ed. Vitelli, Berlin, 1888.

W. D. Ross: references to Ross are to his commentary on the *Physics* (Oxford, 1936) ad loc., unless otherwise stated.

Simplicius, *In Aristotelis Physicorum libros quattuor priores commentaria*, ed. Diels, Berlin, 1882.

M. Untersteiner, *Aristotele, Della filosofia*, Rome, 1963.

J. C. M. van Winden, *Calcidius on matter*, Leiden, 1959.

W. Wieland, *Die aristotelische Physik*, Göttingen, 1962.

E. Zeller, *Aristotle and the earlier Peripatetics*, tr. Costelloe and Muirhead, London, 1897.

GLOSSARY

in accordance with: κατά
account: λόγος
on account of: διά
activity: *see* rational, practical
actuality: ἐνέργεια
affection: πάθος
alteration: ἀλλοίωσις
arrangement: τάξις
automatic: αὐτόματος
beginning: ἀρχή
being: εἶναι
  being of good (185ᵇ21–2), in being (186ᵃ31), etc.: ἀγαθῷ εἶναι, τῷ εἶναι, κτλ.
  what the being would be: τὸ τί ἦν εἶναι
  what is: τὸ ὄν
  precisely what is: (τὸ) ὅπερ ὄν
belong: ὑπάρχειν
cause: αἰτία, αἴτιον
change: κίνησις, μεταβολή
choice: προαίρεσις
  object of choice: προαιρετόν
come to be: γίγνεσθαι
composite: σύνθετος
  composition: σύνθεσις
concur: συμβαίνειν
  by virtue of concurrence: κατὰ συμβεβηκός
continuous: συνεχής
differentiating principle: διαφορά
disposition: διάθεσις
due to: dative of noun
element: στοιχεῖον
end: τέλος
for (preposition, emphatic, *see* 194ᵃ36 and commentary ad loc.): ἕνεκα
form: εἶδος, μορφή
genus: γένος
of, by, in, itself: καθ' αὑτό
kind: γένος
  kind of thing (I. 6): γένος, γένος τοῦ ὄντος
lack: στέρησις
limit: πέρας
  limited: πεπερασμένος
luck: τύχη
matter: ὕλη
movement (193ᵃ30): κίνησις
nature: φύσις
  natural (except at 184ᵃ16): φυσικός
  be by nature such: πεφυκέναι
  student of nature: ὁ φυσικός
opposite: ἐναντίος
  opposition: ἐναντιότης, ἐναντίωσις
  opposed: ἀντικείμενος
outcome of: ἀπό
pass away: φθείρεσθαι
position: θέσις
possibility: δύναμις
  possible (191ᵇ28, 195ᵇ16, 20): κατὰ (τὴν) δύναμιν
practical activity: πρᾶξις
principle: ἀρχή
qualitative change: ἀλλοίωσις
rational activity: πρᾶξις
reality: οὐσία
responsible for: αἴτιος
separable: χωριστός
shape: σχῆμα, but in 193ᵃ30, ᵇ4: μορφή
simply: ἁπλῶς
species: εἶδος
start, starting-point: ἀρχή
state: ἕξις
supervene: συμβαίνειν
tendency: ὁρμή
this thing here: τόδε τι
thought: διάνοια
underlie: ὑποκεῖσθαι
  underlying thing: ὑποκείμενον
universal: καθόλου
the universe: τὸ πᾶν
unlimited: ἄπειρος
work: ἔργον

INDEX

The figures in bold type refer to the text

PRINTED IN GREAT BRITAIN
AT THE UNIVERSITY PRESS, OXFORD
BY VIVIAN RIDLER
PRINTER TO THE UNIVERSITY